Innovation Policy and Canada's Competitiveness

Innovation Policy and Canada's Competitiveness

by Kristian Palda

The Fraser Institute
Vancouver, British Columbia, Canada

Printed in Canada.

Canadian Cataloguing in Publication Data

Palda, Kristian S., 1928 –

Innovation policy and Canada's competitiveness

Includes bibliographical references.
ISBN 0-88975-154-4

1. Technological innovations—Canada. 2. Research, Industrial—Canada. 3. Competition—Canada.
4. Industry and state—Canada. I. Fraser Institute (Vancouver, B.C.) II. Title.
HC120.T4P34 1993   338.971   C93-091763-4

Table of Contents

Preface

IN 1984 THE FRASER INSTITUTE PUBLISHED my *Industrial Innovation: Its Place in the Public Policy Agenda*. When I first started on the present volume, I believed it to become a second, updated edition of *Industrial Innovation*. Yet it turned out to be so different a book, using perhaps only a fifth of the previous material, that it deserved a different title.

I would like to thank Michael Walker and Fred Mannix for their kind and effective support. The monograph benefited from my research association with Petr Hanel of Sherbrooke University and with my two Queen's colleagues, Bohumír Pazderka and Lewis Johnson. Another Queen's colleague, Klaus Stegemann, provided valuable information. Linda Freeman typed the manuscript quite admirably.

The most substantial help came from my son Filip, to whom I dedicate this book.

Kristian Palda

About the author

KRISTIAN PALDA IS PROFESSOR of business at Queen's University, Kingston, Ontario. Born in Prague, Czechoslovakia in 1928, he completed his undergraduate education at Queen's University in 1956 and obtained his M.B.A. and Ph.D. from the Graduate School of Business of the University of Chicago, where he was the recipient of the 1963 Ford Foundation doctoral dissertation prize.

Professor Palda taught business and economics at the Ecole des Hautes Etudes Commerciales in Montreal from 1958 to 1962, when he was appointed assistant and then associate professor at the State University of New York at Buffalo. In 1965 he went to Claremont Graduate School, Claremont, California where he was professor of business economics until 1970. He was appointed to his present position at Queen's University in 1970. He held visiting appointments and has lectured widely in North America and Europe, especially in French-language areas.

Professor Palda's research interests, always coloured by economic analysis, have been devoted to two fields: the examination of advertising effects in commercial and political markets, and issues in technological innovation. He has published seven books touching on these areas, the latest two being *Industrial Innovation* and *The Role of Advertising Agencies in Canada's Service Sector* with the Fraser Institute. His most recent articles on campaign spending and technological innovation appeared in *Journal des économistes* and *Prometheus*, respectively.

In 1987 Professor Palda won a Queen's University prize for excellence in research. In 1991-92 he spent his sabbatical year at the Prague School of Economics.

Introduction

THE FIRST PURPOSE OF THIS VOLUME is to provide adequate information and guidance to understand innovativeness and the policies designed to stimulate it. Ample statistics and references to recent literature are offered so that the reader will be able to reach an independent judgement.

Technological innovativeness is conceived of as both the creation of new products and processes and the receptivity of enterprises to the adoption of new or improved equipment or methods. Just as industrial research and development is often a necessary activity in the creation or reception of new technology, so innovativeness itself is frequently a required ingredient in the achievement of competitiveness.

Yet to this day, most of the governmental industrial policy thrust in Canada, federal and provincial, is aimed primarily at R&D support and to some extent at aid to diffusion. The second goal of this volume is therefore to make it clear that innovativeness is not an activity that operates in a vacuum. Just as any other economic endeavour, innovativeness—and the necessity of staying competitive—demands managerial direction and investment funds as backing.

Yet managerial effort and investment capital will not be forthcoming if what we call "business conditions" are not right. An economy saddled with excessive anti-market rhetoric, regulation, taxation, and subject to wide swings in currency exchange and interest rates will simply not appeal to dynamic managers and venturesome investors. To believe that another tax concession or specific subsidy to this or that R&D project, that yet another pinpoint industrial strategy will open up the horn of plenty is naive. If not naive, then self-serving and rent-seeking. Canada has had for years and still has the most generous R&D-plus-innovation-diffusion tax support system of all leading industrial countries and yet has not progressed an iota in its overall research intensity.

For this reason, and for many others discussed in this book, we are strongly opposed to any *increase* in provincial and federal taxpayer support of industrial research and technology diffusion. Although this book makes an effort to present the subsidization viewpoints, we do not find them persuasive.

So while innovativeness is an essential ingredient of a competitive economy, the basic conditions for its *further* flowering in Canada are not present. It would take a fundamental reversal in governmental fiscal and infrastructure policies, especially on the provincial level, to offer the incentives to innovation which industrial policy is unable to provide.

Kristian Palda

Chapter 1

Innovativeness: The Principal Issues

The government's optimism about technology knows neither program-matic, partisan, nor ideological bounds.[1]

Introduction

TWO MAIN THEMES RUN THROUGH THIS MONOGRAPH. The first is tech-nological innovativeness—its determinants and its repercussions. The second is government policies that sometimes facilitate, but more often impede innovation's genesis and diffusion.

Technological innovation finds its embodiment in new or improved products or processes. To earn the "innovation" designation such products or processes must undergo the test of the market, whether they are successful or not. Without the commercial try-out, we can only speak of "inventions" or improvements. The point of technological innovation is that it either widens the scope of customer choice (new products) or

---

1   Linda Cohen and Roger G. Noll, *The Technology Pork Barrel*, Washington: The Brookings Institute, 1991, p. 1.

lowers the purchase price (new processes), or both.[2] Thus, it enhances the economic well-being of the nation.

But these effects are not easily measurable on the level at which they count most, that of the final consumer. Thus, we tend to concentrate on measuring changes in cost and productivity both among innovating firms and their *commercial* customers, changes consequent upon innovation's creation and *adoption*.[3]

It is clear that the efficiency (or productivity) of commercial customers can be enhanced by the installation of equipment, such as machinery or computers, purchased from innovative suppliers. We call such a purchase an "innovation adoption," which is considered as vital to economic progress as "innovation creation." A firm's, an industry's, an economy's *innovativeness* can thus be envisaged as consisting of both types of activities—the creation as well as the adoption/diffusion of innovation. Both are equally important, but their relative shares will depend on a host of influences, such as the resource base of the economy or access to large markets.

Importance of Innovativeness

A first impression of the estimates of economic progress ascribed to technological innovativeness is shown in a table compiled from the Nobel prize acceptance speech of the American economist Robert

---

2   *Techological* as opposed to *organizational* or *managerial* innovation. The latter may not rely at all on changed product or process configurations. Corporate decentralization in the '20s is an example, though clearly it could not have happened without the '20s automobile and the telephone to facilitate it.

3   A moment's reflection shows that the consumer household is a small factory in which market purchases are combined with the services of domestic appliances and the household's time to produce final, utility-yielding goods. We can therefore equally speak of household productivity, substitution of inputs, etc. and imagine what, for instance, time-saving innovative equipment contributes to consumer welfare. But household activities are not statistically assessed since they are not a part of the monetized economy. We thus have a less clear picture of innovativeness at this level.

Table 1: U.S.A. 1929-1982

Average annual growth of real business output	3.1%=100
Can be attributed to:	
Increased Labour Input	25%
Increased Educational Qualifications of the Average Worker	16%
Increased Capital Input	12%
Improved Allocation of Resources (such as movement of labour from agriculture to manufacturing)	11%
Economies of Scale	11%
Technological Progress	34%
Total	109%
Less such negative factors inhibiting growth as increased government regulation	–9%
Total	100%

Source: Robert M. Solow's 1987 Nobel prize lecture "Growth Theory and After," *AER* (1988) v. 78, no. 3, p. 314, quoting Denison's work (1985).

Solow.[4] Table 1, derived from Denison's estimates—which are based upon Solow's pioneering methods in detecting determinants of economic growth—attributes one-third of U.S. growth in the private sector output between 1929 and 1982 to technological progress. This is the kind of figure which makes politicians salivate when they contemplate the usually dismal economic statistics detectable in their wake. If this sort of golden-egg laying goose could be reared by relatively inexpensive

---

4   Robert Solow, "Growth Theory and After," *American Economic Review* (1988), v. 78, p. 314. E. Denison, *Trends in American Economic Growth, 1929-1982*, Washington: The Brookings Institution, 1985.

industrial policies, then budget-tightening measures could be discarded in favour of taxpayer-financed projects dear to every pressure group in sight.

There are, of course, other ways of measuring the outcome of innovative activity. An example is an appraisal of the contribution of biomedical research to health production. Vehorn et al. attempted to measure the degree to which this type of research lowered mortality in the United States over the years 1930 to 1978. One of their regressions is shown in Box 1. It appears that, during that period, a 1 percent increase in research effort (measured by biomedical PhDs awarded 10 years previously) resulted on average in a 0.05 percent reduction in mortality

---

Box 1:
Effects of Biomedical Research on Mortality, U.S.A. 1930-1978

$$M = 3.75 - 0.17Y - 0.35L + 0.03U + 0.25I$$

        27   31   -3   13

   - 0.05R(10) $N = 49$ $R^2 = 0.91$

     23   All variables statistically significant at 0.01

M=Age-adj. mortality per 1000 population
Y=Real per capita income in 1967 $
Y=P.C. stock of doctors and nurses
U=Unempl. rate, lagged 2 years
I=Work injury rate

R(10) = Biomedical PhD's awarded by US universities,
         lagged 10 years, a proxy variable for biomedical research

Numbers below coefficients indicate the share of the variable's contribution to R^2.

1975 Worth of mortality reduction over 1930-78 = $846B
1975 Worth of biomedical research expenditures = $ 86B
        "Savings" = (846 x 0.20) - 86 = $83 Billion

Source: Charles L. Vehorn, J. Steven Landefeld, Douglas P. Wagner, "Measuring the Contribution of Biomedical Research to the Production of Health," *Research Policy* (1982), pp. 3-13.

rates. Research effort in this estimate accounted for 23 percent of the total of 91 percent explained in the movement of mortality rates.

The economic value of the reduction of mortality in this period was estimated to be worth $846 billion in 1975 dollars, while an independent estimate of biomedical research expenditures gave $86 billion. If we take only 20 percent, rather than 23 percent, as the contribution of biomedical research to mortality reduction, we obtain a net return to such research of $83 billion. This does not capture the improvement in the health status of the living. Clearly, the payoff to this type of research was as impressive as the payoff to technological progress in the Solow calculations.

Measurement Issues

The above mentioned studies are interesting not only because of the importance of the outcome, but also because they are so clearly of a multivariate character. Output growth and mortality decline are found to be only partially, though quantifiably, affected by innovative activities. Other influences, such as those of the stock of medical personnel or of capital, are included and evaluated by standard statistical methodology.[5]

Politicians, lobbyists, and sometimes even scientists, when clamouring for increased support to research, only rarely pay attention to the existence of multiple causes. They prefer to dwell in a single bivariate world where single causes lead to unique outcomes.

These two studies also attempt to uncover whether innovative inputs have desirable impacts. Too often in public debate the attention is focused on only one or the other side of the innovative "function." Korea's economic growth is fast, therefore its R&D outlays are probably high; Phillips' research expenditures are large and increasing, therefore the electronics giants' profit growth must be strong. That the desirable economic outcome necessarily results from innovativeness, or that innovativeness necessarily leads to economic success has, however, no general proof. The connection, or more accurately, the *partial* connection between them must always be documented. This can be painful and difficult.

---

5   The Denison-Solow results are based on regression estimates of production function parameters.

A hint of the inherent difficulties underlying the measurement issue is already implicit in the two studies mentioned. Why did Solow examine the growth of business output and not the growth of the total, that is business plus government output? We do not register productivity changes in the public sector because its output is measured by its inputs, with no changes in value added. Why did Vehorn et al. measure the decline in mortality when new drug therapies not only prolong life, but also make people enjoy better health while living, i.e., lower morbidity, the incidence of diseases? Simply because morbidity is a much more elusive concept than mortality.

Figure 1 offers some clues about what should be taken into account when trying to measure the degree of innovativeness, its determinants, and its consequences. These (and other) measurement aspects, which will be constantly alluded to throughout this volume, are not examined for their academic interest. They are sketched here in a first "fly-by" so that the problems facing adherents of policies to stimulate innovativeness can be glimpsed.

The Input Side

Panel (a) of figure 1 deals with what can be called the input side of innovativeness, as well as with its two constituent parts. Let us consider those two first. Since innovativeness was defined as consisting both of product/process innovation creation and of innovation adoption/diffusion, it is readily apparent that its degree or intensity cannot be measured directly. An index of innovativeness can only be an aggregate of apples and oranges: technological breakthroughs mixed with incremental process improvements and widely differing diffusion rates in a host of markets.[6] The same reservations apply to potential indices of "just" innovation creation or diffusion.

---

6   It will be seen later that similar difficulties beset an index of competitiveness. Nevertheless a start has been made on at least defining and counting innovations. The U.S. Small Business Administration has catalogued 8,074 innovations introduced in the United States in 1982. For an analysis of these data see Zoltan J. Acs and David B. Andretsch, "Innovation in Large and Small Firms," *American Economic Review*, September 1988, pp. 678-690.

Figure 1: A Representation of Determinants (a) and Consequences (b) of Innovativeness

(a)

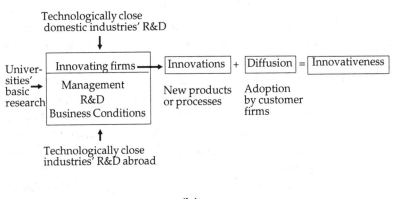

(b)

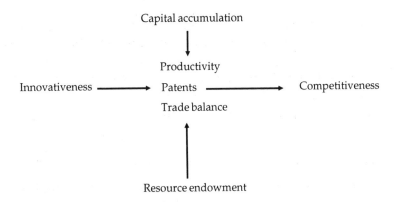

And so it is not surprising that much of the discussion and investigation of innovativeness focuses on the imperfect indicator of research and development expenditures, or research intensity (R&D divided by sales revenue). At the same time the R&D indicator is also considered to be a prime *determinant* of innovation. It thus serves two purposes: as a proxy measure of innovativeness and as an input into the process of innovation. But as an input it is merely a necessary—certainly not a sufficient—contributor to industrial innovation.

Before we look at the other contributing elements let us pay closer attention to R&D. It is only over the last decade that we are obtaining ever stronger econometric evidence about the existence of so-called technological spillovers.[7] Technical knowledge gained through research spills over, without cost, to technologically similar industries. The recipient firms or industries need not be, and often are not, in the same country. As well, industrial R&D also relies on knowledge generated through basic research in universities and research institutes.

As has been repeatedly shown, successful industrial innovation is predominantly the result of skillful management. Management is the bridge between the laboratory and the market or the factory. Management evaluates a market or a process need, entrusts its researchers with meeting it, and guides the production specialists in delivering the new product or installing the new process at acceptable cost. The essence of innovation, states a recent (1990) research project outline of the Science Council of Canada, is the commercialization of technology rather than its mere creation.

Finally, a typical innovation requires a substantial investment. It must be financed not only during its gestation in the lab, but also in the market investigation phase, in the setting up of pilot runs, and production and distribution facilities, in advertising its launching to markets. Such an investment will only be incurred when business conditions—cyclical as well as political—are welcoming.

---

7   Jeffrey Bernstein, "The Structure of Canadian Interindustry R&D Spillovers and the Rates of Return to R&D," *Journal of Industrial Economics* (1989), v. 37, No. 3, pp. 315-328.

What, then, are the implications of panel (a) for the would-be dispensers of taxpayer funds to enhance innovativeness? The most obvious one is that the diffusion/adoption component must also be taken into account. The second implication is that since there is no satisfactory way to measure (no "index" for) either innovation creativity or diffusion, only some very indirect proxies can be pursued as objectives, such as R&D activity.

Yet the objective of increasing R&D effort suffers from two handicaps: not only is R&D but a proxy variable for innovation, it is also only one of several inputs into innovation creation.

We singled out business conditions as being crucial to the investment plans behind innovative products and processes. Such conditions are fundamentally influenced by the macroeconomic as well as "target-neutral" policies of the governments. The encouragement of a high level of savings and capital formation by sound fiscal and monetary policies, the provision of an adequate infrastructure of technical and managerial training, the absence of stultifying regulation are but some examples of what is meant by policies favourable to investment.[8] Clearly such policies are much more exacting than just direct support to R&D.

The input factor "management" is particularly relevant at the level of the individual enterprise. Should only well-managed enterprises be the beneficiaries of taxpayer support to R&D? How do we predict the quality of management during the course of the supported innovation project? Can bureaucrats pick winners?

And if taxpayers are to subsidize one specific industry, they will inevitably contribute—because of spillovers—to the innovativeness of another, possibly foreign, sector. These are some of the issues that anybody considering how to increase innovativeness in a firm, industry, or country must grapple with.

It is not idle, even at this introductory stage, to insist that general framework policies which make for a sound economy, as well as managerial skills (especially in marketing), are as much responsible for industrial innovation as scientific research and development itself. Even

---

8   E.P. Neufeld, senior V.P. of the Royal Bank, before the Senate Subcommittee on Estimates, May 12, 1983.

NABST, the National Advisory Board on Science and Technology to the prime minister, seems to stress only one part of the picture, the technology part, in its 1991 normative scenario for a Canadian industrial policy, as evident in Figure 2.

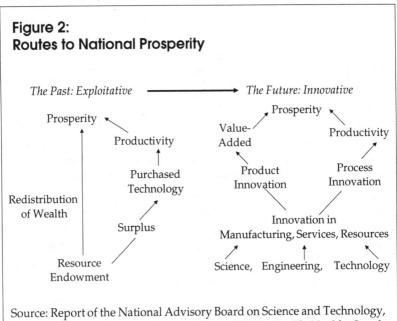

Figure 2:
Routes to National Prosperity

Source: Report of the National Advisory Board on Science and Technology, *Science and Technology, Innovation and National Prosperity: The Need for Canada to Change Course*, Ottawa: April 1991.

The Output Side

Panel (b) of Figure 1 indicates the possible consequences of innovativeness and the *output* proxy measures of it. Again, no single satisfactory measure is available.

Patents, for instance, will not be present in an industrial sector (or in an economy) where innovative activities are mostly of the adoption kind, or where intellectual property cannot be adequately protected by them (eg. mining, petroleum refining). Apart from innovation, the balance of trade in a specific sector, i.e. the difference between exports and imports, will clearly be influenced by a whole host of factors, including, for instance, a sector's natural resource endowment. Total factor (capital plus labour) productivity will in part depend on the amount of capital available and on the training of labour.

Nowadays it is fashionable to subsume a sector's long-term health and prospects under the word "competitiveness." One could therefore also propose competitiveness as an aggregate index of the outcomes of innovativeness. The difficulties of constructing such an index, which typically consists of at least two components—trade performance and productivity growth—are obvious. But since at some very basic level the notion of competitiveness seems to be right and important, we shall pay further attention to it subsequently.

Nevertheless, since the *c* word is now so freely bandied about, not just in connection with a particular sector of industry but also in relation to a whole economy, it is advisable to quote here and now a well known observer of the economic scene, Paul Krugman of MIT:

> Indeed, trade between countries is so much unlike competition between businesses that many economists regard the word "competitiveness," when applied to countries, as so misleading as to be essentially meaningless.[9]

Is Canada Innovative and Competitive?

Having sketched some aspects necessary to the preliminary under-standing of the policy issues regarding innovativeness, we now present some statistics which preoccupy so- called decision makers in this area. A greater in-depth exploration is reserved for subsequent chapters.

The most infamous and notorious indicator of the alleged lack of innovativeness in Canada is the GERD/GDP ratio, or the gross expen-ditures on research and development divided by the gross domestic product. As Figure 3 shows, this presumed indicator of an economy's innovativeness—as signalled exclusively by research and develop-ment—shows Canada to be in the minor leagues, with a ratio of 1.33 percent in 1989.[10]

---

9   Paul A. Krugman, "Myths and Realities of U.S. Competitiveness," *Science*, November 8, 1991, p. 811.

10   Later Canadian figures are virtually unchanged; international figures come

Figure 3: GERD/GDP, Approximate Percentages, 1989

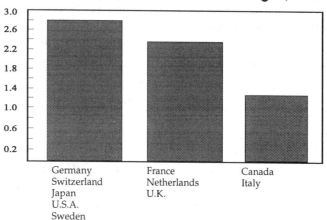

Germany	France	Canada
Switzerland	Netherlands	Italy
Japan	U.K.	
U.S.A.		
Sweden		

Source: Industry, Science and Technology Canada, Selected S&T Statistics, 1991, Ottawa: December 1991.

This is the statistic most frequently used in debate by opposition parties (Progressive Conservatives before 1984, Liberals after, Socialists throughout), by the media, and by many science and technology interest groups to show that Canada lags behind in innovativeness, and so ultimately, in competitiveness. While we shall return at greater length to this very indicator, it is of interest at this point to present—in Figure 4—an intriguing relationship between the size of an economy and its outlay on R&D.

A regression line of "best fit" between the natural logarithms of GERD and GDP appears to have a slope statistically significantly greater than 1 (actually 1.16): the larger an economy, as measured by its domestic product, the higher a proportion of it is spent on research. If this exponential relationship could be explained on economic grounds, policy-makers' efforts to increase the GERD/GDP ratio would seem to lose even more of their justification.[11]

---

with a delay of 3-4 years.

11   The figure is taken from J.A.D. Holbrook, "The Influence of Scale Effects on International Comparisons of R&D Expenditures," *Science and Public Policy*, August 1991, pp. 259-262. The countries are Australia, Austria,

Figure 4: Log of GERD as a Function of Log of GDP 1987, 19 countries

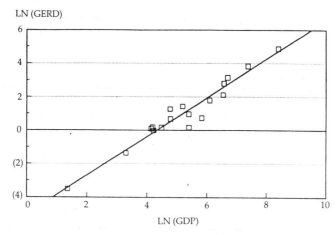

Source: J.A.D. Holbrook, "The Influence of Scale Effects on International Comparisons of R&D Expenditures," *Science and Public Policy*, August 1991.

The next graph (Figure 5) purports to show a deterioration in Canada's competitiveness over the years 1980 to 1990. Competitiveness, just as the GERD/GDP ratio, is conceived of as an economy-wide measure. In this representation, offered by the Canadian Manufacturer's Association in 1990, an index composed equally of unit labour costs, wholesale prices, and international trade is pitted against a similar composite index for the other countries comprising the G-7 (US, UK, Japan, France, Germany, Italy). The aggregative steps taken in compiling the index are breathtaking and are compounded by the element of international comparison. If innovativeness is an important element contributing to competitiveness, then—on this reckoning—our presumably poor showing on one index implies a poor performance on another.

Belgium, Canada, Denmark, Finland, France, Germany, Iceland, Ireland, Italy, Japan, Netherlands, Norway, Spain, Sweden, Turkey, U.K., U.S.A. The economic explanation could be based on the assumption that it is easier to appropriate returns to innovation in a larger jurisdiction. See Bohumir Pazderka, "Determinants of the International Distribution of R&D Expenditures," Canadian Economics Association Meetings, June 1992.

Figure 5: The Canadian Competitiveness Index
1980 = 100

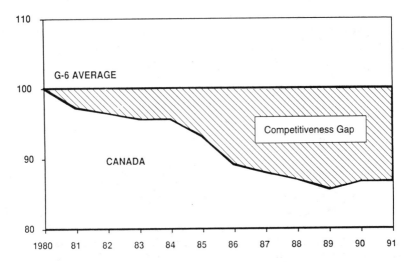

Source: Canadian Manufacturers Association, September 1991 newsletter.

The third measure of alleged shortfall in innovativeness is actually a small component of the previous one, but more sharply focused. It is the share of trade in high-technology products. Again, we shall return to this issue in greater detail later asking, for instance, why the definition of hi-tech does not include the nuclear reactor sector, or why market share is a decisive factor when in fact the whole market grows furiously. As Figure 6 indicates, Canada's share of OECD's exports of the so-called hi-tech products has declined slightly over the 28 years between 1963-1987. (At the same time, medium-research intensive products are doing very well over this period). But this argument must not pass unchallenged, even in this brief exposition. When Northern Telecom buys a component from its plant in Texas, this is registered as a Canadian import. If trade balances were calculated in terms of ownership, it would not. Is a geographical definition superior to one in ownership terms? It is well known in this context that U.S. corporate sourcing practices, that is, U.S. corporations buying from their own plants abroad, account for a substantial part of U.S. trade imbalances with Taiwan, Singapore, and South Korea.

Figure 6: Canada's Share of OECD's Exports by Technology Type, 1963 to 1987

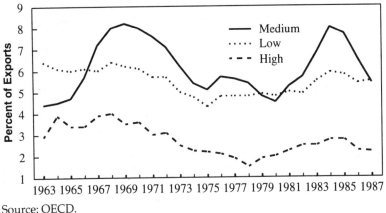

Source: OECD.

Figures 3 (GERD/GDP), 5 (Competitiveness), and 6 (Trade in Hi-Tech) thus express in a condensed way the malaise about Canadian innovativeness that has been part of this country's conventional wisdom for at least two decades.

Direct Government Support for Innovativeness

During the same time, a blanket accusation against successive federal governments states that not enough public funding is being committed to innovative activity in general, and to research and development in particular. The federal taxpayer supports research through various outlets: federal laboratories (such as in agriculture and environment), grants (to universities and industries) and tax abatements to private sector research performers. In a later chapter this topic will be explored at greater lengths. Here we show, in Table 2, perhaps the most important facet of taxpayer's generosity toward research. It is an international comparison of Canada's tax treatment of R&D.[12]

---

12   Jacek Warda, *International Competitiveness of Canadian R&D Incentives: An Update*, Report 55-90, Ottawa: The Conference Board of Canada, June 1990.

Table 2: Comparison of B-Indexes in 10 Countries, 1989 and 1981

Country	1989 (Current Study)		1981 (Previous study)	
	B-index	Rank	B-index	rank
Canada	.657	1	.84	1
United States	.972	5	.95	2
Australia	.703	2	1.01	5
Japan	1.003	7	.98	3
Korea	.805	3	1.01	5
France	.813	4	1.02	6
F.R.G.	1.027	8	1.05	8
Italy	1.033	9	1.03	7
Sweden	1.04	10	.95	2
United Kingdom	1.00	6	1.00	4

Note: This comparison assumes for Canada a Quebec-based corporate income tax system; for the U.S., the California tax system; for Korea, the claiming of double depreciation incentive; and for France, the claiming of the volume-based investment tax credit.

Source: Jacek Warda, *International Competitiveness of Canadian R&D Incentives: An Update*, Report 55-90, Ottawa: The Conference Board of Canada, June 1990.

The table shows a so-called B-index rating of tax incentives available to research-performing corporations. It is defined as after-tax cost of $1 of R&D expenditure, divided by one minus the tax rate, or ATC/(1 – tax rate). Both in 1981 and in 1989 Canada ranked first in generosity, if the corporation was subject to the Quebec taxation regime. (If the B.C. corporate tax system were to apply, the least stimulative in Canada, then Canada's rank would decline to third place, if Ontario's, then to the second). This investigation was carried out by the Conference Board of

Canada and financed by the federal department of Industry, Science and Technology. Its results indicate strongly that at least as far as the tax system (federal and provincial) is concerned, Canada is more than enough competitive.

Summing Up

In this chapter we have described a number of building blocks that should go into the construction of a book dealing with Canada's innovativeness and with policies aimed at improving it.

We stated that technological innovativeness consists of both the creation and the diffusion of new products and processes. We showed two examples—one of economic growth, the other of mortality decline—due to innovativeness. This to illustrate the general beneficial effects of it. Yet we cautioned that it is quite difficult, in most instances, to be precise about the role that innovativeness plays in such a complex phenomenon as economic prosperity. Definitional and measurement problems account for this: R&D is but a proxy measure of the larger concept, for instance, and it is only one factor in the multivariate causation of economic prosperity.

Figure 1 attempts to illustrate the argument and represents, in a simplified manner, the underpinning for this whole monograph. Innovativeness can only be represented indirectly by input (R&D) or output (total factor productivity, etc.) proxies. There are several other important inputs into innovativeness (management, etc.), and other outputs by which it can be assessed in part (patents, etc.). Sometimes several of these proxy outputs are brought together into an index of competitiveness.

Figure 1 led us to preliminary reflections on the usefulness of pinpoint industrial policies toward R&D, on the subsidization of a domestic sector which may contribute to the innovativeness of a rival foreign sector, and on other issues.

Finally, this chapter listed graphs and statistics supporting the primary concerns of Canadian media, politicians, and technology lobbies. These are the GERD/GDP ratio, the competitiveness index, and the hi-tech products trade balance. We noted that these concerns occur despite the most generous R&D taxation climate anywhere.

Looking Ahead

Future chapters will present a coherent view of the state of Canada's innovativeness and of the policies—past, present and proposed—designed to improve it.

The next chapter begins by defining industrial policy and by examining traditional (market failure) and its recent (strategic trade) conceptual foundations. Since much of the debate about industrial strategy centers around technological innovation, it will not be necessary to make a special effort to narrow it down to our own topic.

Having established the theoretical reasons for a public policy debate about innovativeness, we must next deepen our understanding of it. This will be undertaken in chapters 3 and 4. Chapter 3 will look at innovativeness from an economic perspective, while Chapter 4 will add a managerial perspective, especially as regards determinants of innovation other than R&D.

In chapter 5 we shall examine in some detail indicators of Canada's innovative performance and competitiveness. Chapter 6 will discuss industrial policies or governmental support for innovativeness in Canada, mainly under the rubrics of taxation, grants, patent policies, and government-performed research. In chapter 7 some foreign innovation policies and their outcomes will be described, with a view to comparing them with our own. The last chapter will summarize the descriptions and arguments presented in the book and will give some conclusions.

Chapter 2

Industrial Policy

Propositions about where and how market forces work poorly, however, cannot alone carry the policy discussion very far.

R.R. Nelson[1]

Definitions

WHILE THE EXPRESSION "INDUSTRIAL POLICY" or, interchangeably, "industrial strategy" has been around for at least three decades, the idea and its workings go back at least three centuries. In response to a request by the Economic Council of Canada Albert Breton formulated a solid definition of it in 1974. Three centuries before that, the finance minister of Louis XIV, J.B. Colbert, practiced it by promoting and sheltering from competition a number of then hi-tech industries, such as armaments, shipbuilding, and tapestry manufacture.[2]

---

1   Richard R. Nelson, *High-Technology Policies: A Five-Nation Comparison*, Washington, D.C.: American Enterprise Institute, 1984, p. 5.

2   "...he proceeded to reconstruct commerce and industry according to the economic principles known as mercantilism. Utilizing protective tariffs, government control of industry and trade, and navigation laws, Colbert

The Breton definition of industrial strategy starts with a notion that all goods and services, both intermediary and final, can be divided into several broad conceptual classes such that the total output of an economy, Q, in a given period can be represented by the sum of individual sectoral sub-outputs

$$Q = A + R + S + M + ...$$

> where A stands for agriculture, R for resources, S for services and M for manufacturing or "industrial" and the three dots represent other possible output classes.[3]

> For certain problems it could be appropriate to make a breakdown to several types of M-output such that, for instance,

$$M = M1 + M2 + M3 + ...$$

> where M1 is high-technology, M2 medium- and M3 low-technology industrial sectors.

> Suppose now that some socially optimal level of Q or M and of its components can be defined and is designated by asterisks:

$$Q^* = A^* + R^* + S^* + M^* + ...$$

$$\text{or } M^* = M^*1 + M^*2 + M^*3 + ...$$

(To the economist a socially optimal level of output means that perfect competition prevails in all of the economy's markets and all externalities have been internalized—all sectors are at the peak of efficiency and no economic agent can be made better off by changes in the output's composition. To the politician a socially optimal output may not mean anything remotely resembling this abstruse concept. To her it may signify an optimal chance at re-election). Given these concepts the definition proposed by Breton is as follows:

---

organized trading and colonization companies, established model factories and succeeded in extending French industry and trade... Coupled with the extravagance of the king, Colbert's programs drained the French economy." *Funk & Wagnalls New Encyclopedia*, New York: 1973, under Colbert, Jean-Baptiste.

3   Albert Breton, *A Conceptual Basis for an Industrial Strategy*, Ottawa: Economic Council of Canada, 1974.

An industrial strategy is an attempt to reduce the gap assumed to exist between the actual outputs of M-goods and the socially optimal level, between M and M*.

An important aspect of this view of industrial policy derives from its general equilibrium formulation: when the goals of the strategy are defined in terms of the size of a sectoral output, they are also defined in terms of output composition or of the relative size of sectors. An increase of M towards M* implies a change in one or more of the other sectoral outputs, such as A or R.[4]

While Breton's may be the most rigorous definition on record, we often encounter proposals that merely state certain grand objectives,[5] frequently inherently contradictory and usually banal, intended somehow to guide government policies and corporate decision-making.[6]

Before we turn to some illustrations of industrial policy we must add to this section a description of what could be called the international trade version of industrial strategy, namely, *strategic trade policy*. We must do so because in the last half a dozen years this version, or offspring of industrial policy, has become more intensely discussed than its parent.

Perhaps the best way to indicate the intertwining of industrial and trade strategies is to cite from a recent US congressional study:

---

4   "The chronic excess demand for loans created by the administered below-market rate allowed the Ministry of Finance to engage in effective credit rationing to the point of guiding the largest banks to make loans to a specific industry or even to specific firms engaged in the investment race." Kozo Yamamura, "Caveat Emptor: The Industrial Policy of Japan" in Paul R. Krugman (ed.), *Strategic Trade Policy and the New International Economics*, Cambridge, MA: 1986, p. 172.

5   "Therefore the Council recommends that: The government, in consultation with industry, set realistic goals for each industrial sector, work with industry to develop action programs, and publicize the goals and achievements." Report of the Advisory Council on Adjustment, *Adjusting to Win*, Canadian Manufacturers' Association, Toronto: March 1989, p. 80.

6   Gordon Ritchie, "Government Aid to Industry: A Public Sector Perspective," *Canadian Public Administration*, 26 (Spring 1983), pp. 37-46.

Japan and other Asian countries have combined numerous policy tools besides long-term government support for technology R&D to promote selected industries: preferential loans from government banks or banks that follow the government's lead; guaranteed purchases by governmental bodies for home-grown products (e.g., semiconductors for Nippon Telephone & Telegraph, supercomputers for government agencies); government-subsidized leasing companies making guaranteed purchases of advanced equipment and leasing them at preferential rates (e.g. robots, CNC machine tools); formal or informal barriers against imports, removed (or partly removed) only after the domestic industry has become a world-class competitor; strict limits on foreign investment in manufacturing; government negotiations for technology licenses on behalf of industry; government guidance (not always followed) to rationalize industries, scrap over-capacity, and encourage companies to get economies of scale by specializing in certain parts of an industry (e.g., machine tools).

This is industry cum trade policy on a comprehensive scale.[7]

An approximate definition of strategic trade policy can be gleaned from a description of the three constituent words. *Trade* implies that the subject of the policy is movement of goods and services across borders and, increasingly, on a global scale. *Strategic* relates to the international interdependence of policy actions in an oligopolistic environment, that is, in markets dominated by only a few firms. *Policy* means here that governments are the main actors in this game. Each government takes into account some response by foreign firms or governments in calculating its own best course of action.[8]

Finally, we should make a distinction between the industrial and trade policies defined above and "general framework" or "target-neutral" policies. These latter policies do not "target" specific industries (such as hi-tech or bankrupt sectors) or specific activities (such as

---

7   U.S. Congress, Office of Technology Assessment, *Making Things Better: Competing in Manufacturing*, OTA-ITE-433, Washington, D.C.: U.S. Government Printing Office, February 1990, p. 79.

8   Klaus Stegemann, "Policy Rivalry Among Industrial States: What Can We Learn from Models of Strategic Trade Policy," *International Organization*, 43 (Winter 1989), pp. 73-100.

technology); they aim to provide the economy with general conditions favourable to growth. Such policies may imply, if they are to stimulate innovation-based growth, fiscal restraint, conservative taxation and limited intervention.

This is how a recent report of the European Industrial Research Association describes Switzerland's posture toward innovation:

> The main task of the government is to support innovation through the creation, maintenance and support of the infrastructure for education and research as well as to generate a climate favourable to innovation. The latter implies stable and reliable economic conditions. Another important duty of governments is to generate the necessary social acceptance for the role of Switzerland as an industrial nation.[9]

While the rest of this chapter focuses on industrial and trade policies that support innovativeness, we would be ill advised to lose sight of the overriding importance of framework policy. Examining high-technology oriented policies in the five leading industrial countries, Richard Nelson asked the crucial question in this context: to what extent does strength in these (hi-tech) industries flow from general economic strength rather than the other way around?[10]

None of the above attempts at definition should blind us to the fact that the divisions between industrial, trade, or framework policies are merely approximate. In an address to the 1971 meetings of the British Association for the Advancement of Science, its president, Sir Alec Cairncross, opined that innovation is a form of investment and when the economic and social climate is uncongenial to investment, it naturally discourages innovation, too.[11]

---

9   European Industrial Research Association, *Impact of Government Policies on Industrial R&D*, Working Group Report No. 41, Paris: 1990.

10 Richard R. Nelson, *High-Technology Policies, op. cit.*, p. 5.

11 Sir Alec Cairncross, "Government and Innovation" in G.D.N. Worswick (ed.), *Uses of Economics*, Oxford: Basil Blackwell, 1972, p. 19.

Rationales for Public Intervention: Industrial Policy

The interference of governments with market processes dates from time immemorial and is the rule rather than the exception whether in advanced economies, developing, or socialist ones. The justification for such intervention ranges from religious to social to economic concerns. Here we concentrate on the economic rationales underlying government activities that are aimed at influencing economic agents (firms and investors) to act in certain ways that would lead to greater innovativeness.

Presumably these agents need to be influenced, by stick and carrot, to do things which they would not, if left to respond to market pressures. Our specific case assumes that the market detracts firms from some optimal or merely desired level of innovative behaviour such that, for instance, the share of high-technology product output in total manufacturing output is less than it should be:

$$M^*1 \text{ (hi-tech)} > M1 \text{ (hi-tech)}.$$

In that sense markets are said to "fail": there is some "flaw" in the nature of privately organized economic activities.

But how do we know or how can we estimate that

$$M^*1 \neq M1?$$

To this there are several possible answers. One, for instance, looks to comparisons of industrial output categories with foreign countries. Another finds its inspiration in arguments presented to elected (and non-elected) officials by special-interest groups. The one answer that is generally accepted in economic literature, at least as a starting point, explains the likely reasons why unhampered market forces fall short of driving firms, and so industries, to optimal levels of output. The most frequently mentioned reasons, over the last quarter-century, are the following ones.

Market Failure Due to Externalities

Firms producing new technology may not be able to exclude others in society (firms *or* final consumers) from obtaining at lower than full cost the benefits of the new technology. When a new product is sold at a uniform price to all comers (the standard occurrence in our markets and under our legal system), some customers who would have been willing to pay a higher price for the product than the one charged reap what is called a *consumer surplus*. If the product is an input into further production processes (e.g., a computing device sold to banks) this consumer surplus may show up in a measurable way, as an increase in the productivity of the customer industry. Had the innovator been able, through perfect price discrimination—charging a price fitted to each customer's eagerness—to appropriate all of the consumer surplus he created, he would have reaped the "full" returns to his venture.[12]

Imagine now an industry, such as the semiconductor sector, in which patents are difficult to enforce and technical secrets hard to keep since scientists and engineers are highly mobile between individual firms—some of which may be foreign-owned. The protection of the rights of ownership to the intangible knowledge assets is difficult under such circumstances. Because the transit of the technological information is not effected through a voluntary value-for-value exchange for a price, such as must be the case in a market transaction, we speak of *market failure*.

Here, as in the previous case, we are in the presence of an *externality* springing from imperfect or unenforceable property rights. In both instances a positive or beneficial externality is in existence: an unintended outcome of one firm's (innovative) activity redounds to the benefit of customers or competing firms.[13]

Such a beneficial externality conferred upon customers is designated in economic jargon as a *productivity spillover*. When it is received by competitors it is called a *knowledge or information spillover*. While increasing the welfare of the economy, such spillovers obviously detract

---

12   In a well-known realistic example Griliches shows how a patent-protected innovator can increase his profit from $25,000 to $500,000 if he is able to appropriate, by perfect discrimination, all of the consumer surplus. See Zvi Griliches, "Issues in Assessing the Contribution of Research and Development to Productivity Growth," *Bell Journal of Economics*, Spring 1979, p. 98.

13   Dennis Mueller, *Public Choice II*, Cambridge: Cambridge University Press, 1989, ch. 2.

from the income of the innovator who is unable to appropriate fully the economic rewards of his new product or process. Spillover-created *inappropriability* is thus believed to lead to a lower level of innovation activity than would be experienced in markets which do not "fail": entrepreneurs are reluctant to invest as fully in ventures in which their rights are not totally secure.

The logical argument that intellectual property rights are often unenforceable and that this may depress industrial research and associated innovation activities leads then to the recommendation that the gap between private innovation benefit and its private cost be closed by a subsidy. Provided, of course, that the subsidy does not exceed the external benefits created by the innovation in question. We shall return to this basic, all-important argument for research subsidization in a more technical manner in a subsequent chapter. Here we should pause and ask to what extent the general expectation of inappropriability and its consequences is justified.

First then, do we have empirical evidence of spillovers? Research in this area has been accelerating over the last decade and appears to show the existence of spillovers, both between firms within an industry, and between industries—the inter-industry spillovers occurring even on an international scale.

The first to look into productivity spillovers, or the effects of innovative efforts of supply industries upon downstream customers, was Terleckyj as long ago as 1960.[14] He found that most of the productivity increases assignable to innovations could only be detected in customer industries. His findings were repeatedly confirmed both with Canada-wide and Quebec data.[15] Box 2 shows some of the empirical findings. This type of spillover, or external effect, can therefore be taken for granted. While it contributes the necessary underpinning for subsidiza-

---

14   Nestor E. Terleckyj, *Sources of Productivity Advance*, Unpublished Ph.D. dissertation, Columbia University, 1960.

15   H.H. Postner and L. Wesa, *Canadian Productivity Growth*, Ottawa: Economic Council of Canada, 1983. Petr Hanel, J.F. Angers and M. Cloutier, *L'effet des dépenses en R&D sur la croissance de la productivité*, Québec: Ministère de la Science, 1986.

Box 2: Productivity Spillovers in U.S. and Canadian Industries

In 1980 Terleckyj published his regression analysis of the annual rates of change in total factor productivity for the period 1948-66 in 20 U.S. manufacturing industries. Total factor productivity is the dependent variable, R&D/value added, R&D/VA, is the set of some of the independent variables:

	R&D/VA conducted in industry		R&D/VA embodied in purchases		
Regression constant	Privately financed, 1958	Government financed, 1958	PF, 1958	FG, 1958	plus other variables
-6.17	.25*	-.05	.85*	.17	
(1.02)	(2.50)	(.59)	(3.78)	(.37)	
		$R^2 = .69$	(t-ratios in parentheses)		

Only the starred variables are statistically significant; thus only privately-financed R&D had a statistically positive impact productivity growth—and "purchased" R&D has three times the impact of in-house R&D.

In 1983 Postner and Wesa published their regression analysis of the rates of change in (a complicated measure of) productivity for the two sub-periods 1966-71 and 1971-76 in 13 Canadian manufacturing industries:

productivity growth =
    0.63 - 0.05 own intramural R&D + 0.18* received intramural R&D
    + 0.04 own extramural R&D - 0.26* received extramural R&D
    + other variables $R^2 = .69$

Only intramural (carried out within the firm) research embodied in purchased inputs had a positive influence on productivity. Purchased inputs containing extramural (commissioned) R&D had an inexplicable depressing influence on productivity growth.

Sources: Postner and Wesa, see footnote 15. N.E. Terleckyj, "Direct and Indirect Effects of Industrial Research and Development," in J.W. Kendrick and B.N. Vaccara, editors, *New Developments in Productivity Measurement and Analysis*, Chicago, Chicago University Press, 1980.

tion advocates, it is not by itself sufficient. For that, the subsidiser must be able to possess information not easily come by, and must not be influenced by personal considerations. We shall elaborate on this in a later chapter.

Knowledge or information spillovers are the much "hotter" topic, attracting more recent attention. The subsidy justification is here based on the notion that the new technological knowledge generated by one firm seeps out to other firms in its industry (intra-industry spillovers) or in other industries (inter-industry spillovers). But there is no reason to expect such seepage to go one way only. Clearly, if there are externalities in this situation, they are reciprocal: each party both receives from and emits an externality to the other party.[16] While the reception may go unacknowledged by the subsidy candidate, it cannot be neglected by the subsidiser who dispenses taxpayer funds.

Just as in the instance of productivity spillovers, the presence of information spillovers may be detected by productivity or cost studies:[17]

$$c = C(y,w,S)$$

where c is the firm's cost of production, y the vector of outputs, w the vector of factor prices and S a vector of spillover variables. (Among factors or inputs is also own R&D capital). The intra-industry spillovers can be defined as a function of the sum of the R&D capital stocks of all rival firms in the industry. Similarly, the inter-industry spillover-causing variable is defined as the sum of the R&D capital stocks for all other industries in the sample.

This type of a model was estimated by Bernstein with Canadian data from corporations in seven two-digit SIC industries from 1978 to 1981. Spillovers showed up as exerting a statistically significant downward pressure on the costs of the "receiving" corporations. Bernstein's investigations, while subject to econometric reservations, go much further than simply showing spillovers, intra- or interindustry. They also

---

16   Robin W. Boadway and David E. Wildasin, *Public Sector Economics*, 2nd. Edition, Toronto: Little, Brown, 1984, p. 105.

17   Jeffrey I. Bernstein, "Costs of Production, Intra- and Interindustry Spillovers: Canadian Evidence," *Canadian Journal of Economics*, 21, May 1988, pp. 324-47.

lead to estimates of the "wedge" between private and social returns to R&D, that is to the difference between what a typical firm in a given industry can expect to reap from its investment in research and what the returns on this investment are to other firms and industries.[18] In another study Bernstein assigns in pinpoint detail the beneficial (i.e. cost reducing) spillovers received by an industry to the various "spilling" industries.[19] Finally, in certain industries spillovers act to depress (i.e., make less necessary) the receiving firms' investment in R&D, while in others they stimulate it.[20]

Bernstein's and other investigators' work has several intriguing implications. One is that some industries are more senders than receivers of spillover externalities—and perhaps, therefore, ought to be singled out for support. The other is that, as we would expect from an already much older literature on oligopolistic rivalry, firms respond to their competitors' innovative activities by increasing their own.

In that sense, then, while spillovers may exist, they do not lead necessarily to underinvestment in research. Indeed, there is an ancient and perhaps ill-substantiated grudge against the pharmaceutical industry which maintains that there is too much "inventing around and so too much R&D investment."[21]

Furthermore, we now have evidence of cross-border spillovers between industries. Mohnen finds that foreign (US, Japan, France, Germany, UK) R&D spillovers have played a major role both in the growth and in the slowdown of total factor productivity in Canadian manufacturing between 1969 and 1982. Their cost-reducing effect out-

18   A lengthier discussion of social rates of return is found in Chapter 4.

19   Bernstein, "The Structure of Canadian Inter-industry Spillovers, and the Rates of Return to R&D," *Journal of Industrial Economics*, 37, March 1989, pp. 315-28.

20   Irene Henriques, *Four Essays on Research and Development and Spillovers*, unpublished doctoral dissertation, Queen's University, Kingston, Ontario, September 1990.

21   W. Duncan Reekie, "Price and Quality Competition in the United States Drug Industry," *Journal of Industrial Economics*, 26, March 1978, pp. 223-37.

weighs the own-R&D expenditure effect.[22] They tend to stimulate—are complementary to—Canadian R&D outlays.

If there is a substantial spillover to Canada from abroad, it is logical to expect some technological spillovers from Canadian industries outward. The basic argument for subsidization of innovative activity is the inappropriability of full economic returns to it by the firm due to spillovers outward. But if much of the spilling consumer surplus flows to foreign customers, the claim on the purse of the Canadian taxpayer is weakened.[23]

Lastly, there is still some scepticism about the measurement of spillovers and about their intensity. One of the leading US econometricians, Zvi Griliches, opines that:[24]

> Recently there have been several interesting attempts to capture the impact of such spillovers and to model their spread but they have not yielded a convincing estimate of their global impact. Our current understanding of this whole process is still seriously flawed. Moreover, without a major revision and extension of the national income accounts and the development of new data and methods for tracing the flow of ideas from one sector to another, researchers are unlikely to do much better in the near future.

The intensity or pervasiveness of spillovers of new technological knowledge is doubted by some prominent students of the subject. They point out that the transfer and utilization of technical information is a

---

22 Pierre Mohnen, *The Impact of Foreign R&D on Canadian Manufacturing Total Factor Productivity Growth*, Centre de recherche sur les politiques économiques, Université de Québec à Montréal, July 1990.

23 "It appears likely that the new international economic environment has reduced the ability of any one national economy to appropriate the economic returns from basic research performed within its boundaries." David C. Mowery and Nathan Rosenberg, *Technology and the Pursuit of Economic Growth*, Cambridge: Cambridge University Press, 1989, p. 292.

24 Zvi Griliches, "Productivity Puzzles and R&D: Another Nonexplanation," *Journal of Economic Perspectives*, v. 2, Fall 1988, 18-19. See also David M. Levy and Nestor E. Terleckyj, "The Problem of Identifying Returns to R&D in an Industry," *Managerial and Decision Economics*, Special Issue, 1989, pp. 43-49.

costly undertaking and one that requires a technological receptivity that is based on adequate research-based knowledge of the relevant subject area. Thus Cohen and Levinthal: "...our analysis suggests a lesson for technology policy...it implies that the negative incentive effects of spillovers and, thus, the benefits of policies designed to mitigate these effects, may not be as great as supposed."[25]

We end up by acknowledging that the implications of information spillovers, should their widespread existence be confirmed, are far-reaching. Nonetheless, we consider them, as regards industrial policy intervention, quite ambiguous.

Market Failure Due to Other Factors

Higher *differential risk* or *uncertainty* in an industrial subsector may arise for several reasons, such as the absence of an insurance market or fewer opportunities for portfolio diversification. The alleged comparative (to other subsectors) affliction of the hi-tech manufacturing subsector could, in principle, be alleviated by public underwriting of risky projects or by direct investment in venture stocks.

The presence of *competitive imperfections* in the form of cartels, monopolies—whether due to increasing returns to scale or not—and oligopolies, or of similar phenomena on the buyers' side, leads to inefficient market outcomes. Where prices do not indicate relative scarcities, market forces will not allocate resources efficiently and output level will be below optimum. However, it is often maintained that the existence of monopoly (and what are patent rights if not a temporary monopoly) or at least its promise, favours innovative behaviour. In general, monopoly has an ancient claim on public intervention. Usually, however, this type of intervention goes under the name of competition, rather than industrial policy. It is, of course, not only possible but quite likely that the source of much monopolization is government policy past or present, whether it encourages a cartel's formation (e.g. in uranium) or protects its existence (e.g., marketing boards) or indeed launches a monopolistic enterprise itself (Canada Post Corporation).

---

25   W.M. Cohen and D.A. Levinthal, "Innovation and Learning: The Two Faces of R&D," *Economic Journal*, 1989, p. 594.

Public policy itself, then, may be an "outside of the market" force which leads to "market failure." It need not, of course, result in monopoly, but rather more generally in the distortion of the sector's output level. A clear example is the 1980 National Energy Program (NEP) of the federal government which pursued the two undoubtedly inconsistent goals of Canadianization and self-sufficiency.

Finally, it is sometimes alleged that the minimum efficient scale (MES) of innovative operations, such as laboratories, is so large that only one or a few enterprises can afford to undertake them.

Noll provides a masterful though somewhat aging exposé of the theoretical issues and of the empirical importance of these types of market failure.[26] Bletschacher and Klodt, meanwhile, offer a similar, though much broader up-to-date survey.[27]

Rationales for Public Intervention:
Strategic Trade Policy

We have noted that an active, innovative behaviour by one firm or industry may stimulate, unwittingly, a similar increase in such behaviour in competing firms or industries, at home or abroad. But to talk of spillovers in such a case is somewhat awkward. Better perhaps to refer to rivalry, or strategic response.

Strategic trade policy justifications start from the observation, now widely accepted, that a country's comparative advantage in trade can change as a consequence of private economic activity or even of government policy.[28] Not just natural ability or factor proportions, but entrepreneurial initiative using innovative products or processes to be first to market and to exploit learning curve economies appears to confer

---

26 Roger Noll, "Government Policy and Technological Innovation," in K.A. Stroetmann (ed.), *Innovation, Economic Change and Technology Policies,* Basel: Birkhauser, 1977.

27 Georg Bletschacher and Henning Klodt, "Braucht Europa eine neue Industriepolitik," Kieler Discussion Paper 177, Institut fur Welwirtschaft, December 1991.

28 These paragraphs rely heavily on Stegemann, *op. cit.* footnote 8.

often lasting advantage as well. Similarly, governmental support to technical education can enhance the natural ability of a country's inhabitants.

Two models of strategic trade policy demonstrate the possibility that a government can improve national welfare by "shifting profits" from foreign to domestic firms. The first, originally proposed by Brander and Spencer in 1981, is nothing else but a game played in "third" countries by duopolists, each originating in a different country.[29] A particular country's government assures its exporter, whom it wishes to set up as a "Stackelberg leader," of subsidies supporting its sales in third countries, where it is in competition with the other world-industry duopolist.[30] "Stackelberg leadership" shifts profits from the "follower" to the leader and so to the leader's country.[31] Why would, however, the Stackelberg duopolist not be willing to assume the burden of temporarily lower prices in his export markets given the vast profits reckoning from his ultimate leadership?

The export subsidy granted by the government presumably establishes a *credibility* that even if duopolist A were to be met by B's price reductions he would persist and would not retreat from his attempt at leadership. Thus credibility emerges as the only reason for governmental strategic trade policy; when two opposing governments get into action, losses are likely to persist on both sides.[32]

The second model upon which a rationale for trade policy can rest in principle was proposed by Krugman.[33] It is the familiar infant-indus-

---

29   A more closely fitting reference here dates from 1985: James A. Brander and Barbara J. Spencer, "Export Subsidies and International Market Share Rivalry," *Journal of International Economics*, 18, February 1985, pp. 83-100.

30   The most frequent example offered is that of Boeing versus Airbus.

31   A pithy description of game theories and Stackelberg outcomes is given in Jack Hirschleifer, *Price Theory and Applications*, 4th ed., Englewood Cliffs, N.J.: Prentice Hall, 1988, ch. 10.

32   Despite booming sales of the Airbus, the loans subsidizing it appear nowhere near their payback, according to Bletschacher and Klodt, *op. cit.*

33   Paul R. Krugman, "New Theories of Trade Among Industrial Countries," *American Economic Review*, v. 73, May 1983, pp. 343-7.

try argument for initial government protection, enriched by a strategic dimension. If a duopolist is given protection in his domestic market, he receives a scale of production advantage—which may be of the learning curve type—over a foreign rival. This scale advantage then leads to lower costs and so to higher market shares even in third, unprotected markets. By the same token the rival duopolist's costs increase due to lower sales and production. The strategic reason for governmental intervention is that of credibility, as in the preceding model: it would not be credible for a duopolist to attempt profitable expansion of export sales on his own.

In a review of these newly-minted theories of international trade Stegemann puts the important question: do models of strategic trade policy provide guidance to government action?

In Stegemann's opinion the revolution attempting to change the traditional free-trade non-interventionist thinking of the economics profession is ebbing rapidly. Also crumbling with it are the theoretical and empirical supports for strategic trade policy:

> A combination of reasons seems to have been responsible: the realization that the apparent policy implications of models of strategic trade policy are highly sensitive to changes in the special assumptions of these models; the difficulty of identifying real-world situations in which the special assumptions apply; the recognition that the costs of implementing strategic trade policies might easily exceed the benefits, even if appropriate "target" industries could be identified; and the apprehension that economic theory was becoming a supplier of intellectual ammunition for powerful forces that favour protection of particular sectors for the "wrong" reasons.[34]

A less pronounced, but still sceptical view of the efficacy of strategic trade policies, is offered in a later paper by Krugman.[35]

It will suffice to single out two instances to illustrate this list of caveats. One of the assumptions underlying profit shifting is that the domestic firm selected for government support is domestically *owned*.

---

34   Stegemann, *op. cit.*, p. 90.

35   Paul A. Krugman, "Myths and Realities of U.S. Competitiveness," *Science*, November 8, 1991, pp. 811-815.

As the intertwining of ownership across borders increases and joint ventures proliferate, particularly in risky hi-tech sectors which are the usual target for intervention, this assumption becomes increasingly tenuous.

The other shortcoming which merits mention is the partial equilibrium perspective inherent in governmental support to a single industry. As was stressed in the discussion of the Breton definition of industrial policy, stimulating the output of one sector by public intervention will generally have a negative effect on the output of other sectors. The implication of the one-industry support was tested by Dixit and Grossman who modelled several oligopolistic industries dependent on a common resource in fixed supply, such as scientists.[36] When trade intervention attempts to shift profits by stimulating a domestic firm to increase output, it reduces the profitability of the other industries, leading them to contract.

Stegemann concludes his survey of the strategic trade policy arena by stating that economists will continue to resist the popular presumption that a policy enhances national welfare if it raises the market share of domestic producers at the expense of foreign firms.[37] They will keep throwing cold water on mercantilist ideas and will lay bare the group interests served by a particular policy at the expense of other groups in society.

As Grossman put it recently in an important paper, ". . . strategic interventions seek gain at the expense of trade partners, and so invite retaliation . . . but the ultimate goal ought to be a co-operative outcome in which all parties desist from pursuit of strategic gains."[38] Yet, as Stegemann states, the notion of the state having the responsibility to "shape" the comparative advantages of its industry in competition with

---

36 Avinash K. Dixit and Gene M. Grossman, "Targeted Export Promotion with Several Oligopolistic Industries," *Journal of International Economics*, 21, November 1986, pp. 233-49.

37 Stegemann, *op. cit.*, p. 99.

38 Gene M. Grossman, "Promoting New Industrial Activities: A Survey of Recent Arguments and Evidence," *Paris: OECD Economic Studies*, No. 14, Spring 1990, p. 19.

other states is politically powerful. An indication of how powerful it is will be furnished in the next section.

The Debate About and the State of Industrial Policy in Canada

The everlasting debate about this topic in media and in political forums testifies to the strength of the idea that the state should intervene in a pinpoint manner in market processes in order to enhance this or that firm's or industry's competitiveness.[39] And so does the imposition of actual industrial policies upon large sectors of the economy despite their frequent and manifest failure.

One of the more animated recent debates about industrial policy took place in Ontario. The Liberal provincial government of David Peterson, now consigned to the dustbin of history, was very much attuned to the industrial strategy propositions of Harvard's Robert Reich, the liberal economic strategist behind the unsuccessful presidential candidate, Michael Dukakis, and the more successful Bill Clinton.

In November 1987 during the annual first ministers' conference, Premier Peterson proposed "that in order to increase productivity and international competitiveness Canadian governments should provide leadership in committing (sic) to R&D as a national priority.[40] This commitment can best be expressed in the form of a national R&D target .... This action plan provides a policy planning framework to achieve a 2.5 percent R&D target (of gross domestic product) within ten years." The two principal reasons given for increased government support of private sector research are spillovers of the information kind and exceptional riskiness.[41]

---

39 It is not easy and probably not fruitful to go on maintaining a distinction between industrial and strategic trade policies. The distinction served its purpose in uncovering their intellectual roots.

40 Government of Ontario, *A Commitment to Research and Development: An Action Plan*, 2nd ed., Toronto, January 27, 1988, executive summary.

41 *Op. cit.*, p. 15.

This proposal and a budgetary commitment of $1 billion[42] to a so-called Technology Fund was in part the result of deliberations of the so-called Premier's Council, a "multipartite" advisory body chaired by the premier and composed of certain cabinet ministers and "leaders" of business, labour and academic communities. The Council, established in April 1986, was given the mandate to "steer Ontario into the forefront of economic leadership and technological innovation." Sometime in 1989 (the three volumes are not dated) the Premier's Council issued a massive report calling for large-scale government intervention within the scope of Ontario's provincial powers.[43] This report represents one of the latest and probably the most detailed advocacy of industrial policies in Canada ever. It is also a compendium of not necessarily unbiased bits of information about industrial policies and assistance abroad and in Canada.[44]

The industrial policy and trade strategy recommended by the Council in its *Competing in the New Global Economy* follows what are by now standard lines in this "business," though perhaps with somewhat better insight:

---

42   As is usual in such commitments, the funds were to be doled out over a period of many years. On the assumption that $200 million was to be paid out each year and discounting at a modest 12 percent, the present value of the billion shrinks to about $750 million. Furthermore, the Technology Fund, announced when the Premier's Council was created, took half of its budget from existing funds. Also, more than a third of the money is not hard cash but tax credits for companies doing research and development (*Globe and Mail*, January 14, 1991, p. A6). The (mid-?) 1990 annual report of the Ontario Technology Fund lists as actual cumulative expenditures a total of $155 million, covering the fiscal years 1986/87 to and including 1989/90.

43   *Competing in the New Global Economy*, Toronto, 3 volumes.

44   One year later, in January 1990, the report was followed by a much-trumpeted conference in Toronto sponsored by the Council, on the theme of global competitiveness. Or rather, more precisely, on how to devise a strategy for Canada's and Ontario's economic future in a globally competitive world. (*Globe and Mail*, January 16 and 17, 1990, 32 both).

1. Assisting restructuring in the industries by

    a) offering potential investors in mid-size, Ontario-based, *export* -oriented firms tax initiatives in new share issues; the firms to be on *a government-approved list*

    b) helping in worker adjustment

2. Investing in high growth and emerging industries

    a) yet more R&D tax incentives

    b) a strategic procurement plan for Ontario government and Ontario Hydro

    c) a risk-sharing fund for export-oriented firms

    d) refocusing the (public-fund-dispensing) Ontario Development Corporation to provide assistance only to businesses in manufacturing and exporting sectors

3. Improving the entrepreneurial climate for start-up companies in the same sectors

4. Meeting the science and technology imperative by

    a) redirecting government research to industry

    b) the establishment of seven Centres of Excellence, comprised of university and industry participants, to encourage transfer and diffusion of technology into industry.

    And so on ...

These proposed policies are listed here merely as an illustration of typical proposed industrial strategy initiatives.[45] Only 4b, the establishment of university-industry Centres of Excellence (materials research, information technology, groundwater research etc.) has been achieved, at the cost of $200 million over five years. The general evanescence of industrial strategy proposals and the impermanence of actual policy implementations in Canada—and elsewhere—is due to the similarly impermanent nature of democratically elected

---

45   Note the persistent insistence on export orientation as a requirement of eligibility. This is probably the only avenue toward trade policy that a constitutionally limited provincial government can open.

governments.[46] There is generally *less than meets the eye* in industrial policies, proposed and actual. It is thus more fun and probably more to the point to argue about them than to describe them. It is therefore our forecast that the Ontario NDP government's industrial policy, unveiled in July 1992, will wither on the vine as well.[47]

For a fairly thorough description of some industrial policies in the USA, Germany, Sweden, France and Japan, and a good listing of Canada's and Quebec's innovation-oriented initiatives, *Competing* is a briefly useful compendium. A more global perspective is offered in the OECD's yearly *Industrial Policy in OECD* countries. For a truly analytical and objective examination of such policies abroad one needs to go to academic publications. This will be done in a later chapter.

But before we close this section we do offer a summary description of the most directly relevant *federal* subsidy or grant policies vis-à-vis innovation, circa 1990. The reader should remember, however, that much of this description will be, or is by now, out of date. To describe, analyze, or criticize innovation policies in Canada is to shoot at a moving target or to play with a kaleidoscope.

The n-th time freshly reorganized and renamed Ministry of Industry, Science and Technology Canada is "charged with ensuring Canada's international competitiveness through a strong, continuing integration of scientific, technological and industrial strategies and activities."[48] Since about 1987 most of the emphasis appears to be given to S(cience) and T(echnology). The ministry coordinates all of the government's S&T activities, it has an assistant deputy minister for

---

46   As an example, the Ontario government introduced a bill in 1988 for a so-called R&D tax super-allowance. The bill died with the demise, in the summer of 1990, of the Peterson administration. It was reintroduced in the legislature in December 1990. The Council on Technology is being recreated and its mandate broadened under the name of the Council on the Economy and Quality of Life, according to the *Globe and Mail* of January 24, 1991, p. A6.

47   *Globe and Mail*, July 29, 1992, p. 2.

48   This and subsequent information is garnered from Department of Finance, *Canada Budget Estimates*, Part 3, 1990-91, Ottawa, February (?), 1990.

"science advocacy" (e.g. National S&T weeks) and, of course, it hands out subsidies to an alphabet soup of S&T-related undertakings.

Table 3 gives some figures about such grants. The second largest subsidy, of about $45 million in fiscal year 1990/91, is to the IRDP (Industrial and Regional Development Program) action which is now wound down. While this program had a substantial component of innovation support, it suffered, at least initially, from some confusion. This is illustrated in Box 3.

By far the largest recipient of R&D (and other-type) grants is DIPP, the Defence Industry Productivity Program, with over $235 million in fiscal year 1990/91. It funds, among other things, research on aircraft engines at Pratt & Whitney in Quebec. Since it is primarily oriented toward exports, it lacks the "consumer surplus" justification: the spill-over benefits to customers go abroad.

If there is any major tendency to be discerned in the helter-skelter collection of industrial mission strategies carried out by ISTC, it is away from pinpoint help to individual companies' R&D. It is oriented more toward diffusion and adoption (TOP—technology outreach programs, InnovAction—Centres of Excellence). It also finances, along vaguely Japanese lines, "precompetitive" research carried out by industry research consortia (strategic technologies programming, STP).

On the whole, however, basically, fundamentally—no matter how we put it—the overall target of Canadian and provincial innovation policies is the support of R&D. This is the tenor of most political pronouncements and the effect of most "strategies." In this context it is important to note that there is a major shift, on the federal scene, away from direct subsidy to business (most of which is in the control of ISTC) to tax credits to stimulate R&D outlays. This highly relevant observation will be documented and discussed at greater length in a subsequent chapter.

This concentration of attention on R&D is almost certainly a wrong posture on the part of the elected representatives. Let us quote from a recent, heavily data-based article in the authoritative *Research Policy*:

> The above characteristics of R&D businesses can lead to policy implications. The most important implication for policy decisions is the demonstration that R&D is not an isolated decision. Heavy R&D spending is usually associated with a suitable competitive position, with related investments in marketing

Table 3: Selected Grants by Industry, Science, Technology Canada

Dollars	Estimates 1990-91	Forecast 1989-90
Grants to outstanding students pursuing undergraduate degrees in natural sciences, engineering and related disciplines in accordance with the Canada Scholarships Program	15,000,000	10,000,000
Grants to the Canadian Institute for Advanced Research to match private sector contributions to this maximum level	2,000,000	2,000,000
Contributions under the Defence Industry Productivity Program	235,538,000	304,400,000
Contributions to Strategic Technologies	16,800,000	1,600,000
Contributions to Saskatchewan Communications Advanced Network	4,100,000	4,080,000
Contributions under the Microelectronics and Systems Development Program	12,200,000	4,859,000
Contributions for the Advanced Train Control System	4,800,000	4,000,000
Contributions under the Technology Outreach Program and the Technology Opportunities in Europe Program	18,200,000	14,274,000
Contributions under the Industrial and Regional Development Act and outstanding commitments under discontinued predecessor programs	45,183,000	50,765,000
Contributions to the Advanced Manufacturing Technology Application Program	2,000,000	1,553,000

Source: Dept. of Finance, *Canada Budget Estimates, Part 3, 1990-1991,* Ottawa: February(?) 1990, 2-66 to 2-67.

Box 3: IRDP, or How a Government Policy is Implemented

The confusion contained in industrial policy programs that lack a clear economic rationale brings officials charged with implementing them to operate along the lines of a so-called garbage can decision model. In this model an opportunity for choice is viewed as a garbage can into which problems, solutions and decision-makers are dumped as they become available. But goals and choices are not necessarily connected. Yet program administrators will, from the onset, use criteria which meet their desire for safety and familiarity.

Atkinson and Powers used data emanating from 1983-1984 grants under the then newly established IRDP (industrial and regional development program) to find out how the subsidy system worked. This is their second most comprehensive regression:

Generosity = Constant − 4.55 Tier + 5.06 New facility
+ 9.15 Innovation − 6.06 Project size (ln) − 3.79 Food & Beverage
$R2adj. = 0.29$
N = 359 projects
All coefficients significant at 5%

The dependent variable is the assistance offered to applicants as a % of the maximum allowable. A tier is the classification of one of the 260 census districts by "need" (unemployment, income, financial fiscal capacity)—there are four of these, from 1 the wealthiest to 4 the poorest. Project size is measured by maximum government liability (eligible project costs for subsidy times max. sharing proportion).

The results indicate that

- Contrary to expectations, the more economically developed census districts received a higher proportion of funds (for each step up the tier, the grant was 4.5% less generous);
- Projects containing a proposal for a new facility received, on average 5% more generous allocations;
- Projects with an innovative element received 9.1% more generous treatment;
- Projects entailing a larger government liability were less generously received (since variable is in logarithms, a straightforward percentage cannot be given);

Box 3 (continued)

- Applicant firms in the food and beverage sector were treated less generously, by 3.5%.

The authors judge that the administrators, who were well familiar with preceding, overwhelmingly innovation-oriented programs, felt more at ease with financing projects leaning that way, or at least containing new equipment. They were unfamiliar with and perhaps unsympathetic to the regional development component (tier) and naturally risk-averse to larger projects (tier). No explanation for the result in the food and beverage sector.

Source: Michael M. Atkinson and Richard A. Powers, "Inside the Industrial Policy Garbage Can: Selective Subsidies to Business in Canada," *Canadian Public Policy*, 13, June 1987, pp. 208-217.

---

and production, and with policies affecting the product line, the human resource function and export activity.[49]

Recall here Figure 1 which stressed the complementary nature of "management" and research, business conditions and research, and so on. Such complementary behaviour is now widely recognized in professional, bureaucratic, consulting and academic circles. Advocates of industrial innovation policies, such as the Ontario Premier's Council, propose therefore a much wider net of support to prospective "winners," stretching from credit facilities, to specialized labour training, to subsidies and tax breaks. But with such ambitious and wide-ranging interventions proposed, the *implementation* of prescribed industrial policies comes to the fore.

---

49 J. Zif, D. McCarthy, A. Israeli, "Characteristics of Businesses with High R&D Investment," *Research Policy*, 19, 1990, pp. 435-45.

The Difficulty of Implementation

The fundamental reason for directing taxpayer generosity toward innovative activities in the private sector—for undertaking an industrial innovation policy—is the presumed existence of externalities. As was explained, the assumption is that the social product of innovation is greater than the private one. The difference between the two, it is thought, accounts for the less than socially desirable level of private investment in innovation (read: R&D). The gap should be bridged by some form of subsidy and so overcome the "inappropriability" deterrent.

The gap must be measured, however, or at least an effort made to estimate it. The gap is largest when it comes to basic research, smallest in relation to development (the D in R&D). In Canada, it will be shown later, federal support to industry does not show a systematic preference for this ordering.

While it may be the case that it takes more than forty years for the simple economic idea regarding appropriability to make its way into the decision bunkers of hi-tech subsidizers, we should look at the even simpler notion of "picking winners." This seems to be a substantial part of any industrial policy, and so of the innovation kind as well.

The idea is to put "your money," that is to say funds extracted from the unfortunate taxpayers, on firms or industrial sectors that promise to contribute to the growth of the economy and employment, but cannot do so without assistance. These may be "threshold firms," hi-tech firms, export-led sectors, whatever. The essential issue is to forecast which ones will be winners.

For this forecast government agencies will need to possess quite detailed information about how the economy works, what foreign competition will be up to, and how the targeted firm performs. Criticizing a 1982 Science Council of Canada proposed initiative to subsidize "threshold firms," that is firms about to take off successfully, Watson pointed out that if the Council has heard about such firms, so has the stock market.[50] And if the stock market, with its strong profit motive,

---

50   William G. Watson, "It's Still Not Time for an Industrial Strategy," *Canadian Public Policy*, 10, No. 2, p. 207; Guy Steed, *Threshold Firms: Backing Canada's*

has not received such information, why would we expect government bureaucracy to have unearthed it? As Watson further pointed out, the Science Council's study also revealed that several of the key threshold firms were on—or through—the threshold of receivership (in 1982).

The difficulty of choosing winners is forcefully illustrated in the Ontario Premier's Council proposal to back threshold companies.[51] The Council reveals a list of such companies in a table reproduced here as Table 4. The table lists Ontario Threshold Companies in 1987. On our count, by the end of 1990 at least 9 of the 25 companies had either gone into receivership, or had experienced grave difficulties, or had been sold to foreign interests.

The failure to pick winners, to establish successful national champions or to see technology-based megaprojects to completion is also well-documented abroad and will be discussed later.[52] As Richard Nelson, one of the best known American economists in this field puts it:

> A policy whereby government officials themselves try to identify projects that will be winners in a commercial market competition is always seductive, but the evidence, from our studies and others, suggests that such strategy is to be avoided.[53]

And Davis *et. al.*, in a closely reasoned analysis of U.K. state aid to industry, provide statistical documentation of the divergence between official industrial policy objectives and those actually pursued.[54] In regard to R&D industrial subsidies it would appear that the principal policy goal was to "pick winners," that is, to give aid to those industries

---

Winners, Ottawa: Science Council of Canada, 1983.

51   Ontario Premier's Council, *op. cit.*, volume 1, ch. 6.

52   Dirk de Vos, *Governments and Microelectronics: The European Experience*, Ottawa: Science Council of Canada Background Study No. 49, March 1983 and Henning Klodt, *Wettlauf um die Zukunft*, Tuebingen: J.C.B. Mohr, 1987.

53   Richard R. Nelson and Richard N. Langlois, "Industrial Innovation Policy: Lessons from American History," *Science*, February 1983, p. 18.

54   Howard Davis, Dorward, Drive and Topple, "State Aid and Industrial Characteristics," *Applied Economics*, December 1980, pp. 413-28.

Table 4: Examples of Ontario Threshold Companies 1987 ($ Millions)

Company	Exports from Ontario	Total World Sales	Country of Ownership	Product Lines
Magna[a]	$454	$690	Can.	Auto Parts
Mitel	385	453	U.K.	Telecom Equipment, Integrated Circuits, Switchboards
De Havilland[a]	242	300	U.S.	Commuter Aircraft and Assemblies
AG Simpson	190	280	Can.	Auto Metal Stampings
Spar[b]	109	191	Can.	Satellite, Communications, Aviation Products
Husky	102	113	Can.	Plastic Molding Machinery
Menasco	79	85	U.S.	Aerospace Components— Flight Controls, Landing Gear
VME Equipment[a]	76	80	U.S.	Bulldozers, Trucks
Timberjack	75	142	U.S.	Logging Equipment
Long Manufacturing	74	78	Can.	Radiators—Auto, Truck Tractors and Oil Coolers
Linamar	68	80	Can.	Machine Castings
Woodbridge Foam	64	287	Can.	Automotive Foam and Plastic Molded Parts
Fleet Aerospace[b]	64	92	Can.	Aircraft Components, Antennae
Electrohome[b]	62	161	Can.	Video Displays, Printed Circuit Boards
Cognos	57	68	Can.	Software
Leigh Instruments	27	53	Can.	Control Systems, Recorders
Devtek	26	104	Can.	Aerospace, Defence and Electronics

Table 4: (continued)

Company	Exports from Ontario	Total World Sales	Country of Owner- ship	Product Lines
Fisher Gauge	23	40	Can.	Mini Zinc Die Castings and Machinery
Navtel[a]	22	25	Can.	Data Communications Test Equipment
Lumonics[b]	13	65	Can.	Lasers
RBW Graphics	10	61	Can.	Books, Directories, Catalogs, Publications
GEAC	8	64	Can.	Computer Hardware
Gandalf	8	130	Can.	Computer Data Communications Equipment
CAE Electronics	N/A	486	Can.	Aerospace, Electronics, Auto Parts
Galtaco[b]	N/A	94	Can.	Castings, Auto Parts

[a] 1987 Data not available: 1985 Data used.
[b] 1987 Data not available: 1986 Data used.
N/A = Data not available.
Source: Ontario Premier's Council, op.cit., volume 1, p. 157.

already favoured by the market as indicated by profitability and employment. The implicit objective revealed by the correlational analysis of disbursements and industry characteristics in 33 sectors receiving research aid in 1975 was the support of relatively unprofitable and high unemployment industries—an objective appearing to fly in the face of judgement by the market mechanism.

The Difficulty with Motivation

We have already intimated, with the help of Box 2, that the objectives of implementers of industrial policy may not match the objectives of the politicians. Clearly the objectives of politicians may not, in turn, match the interests of the economy as a whole.And so the remedies suggested for market failure, and their implementation, may lead to *nonmarket failure*. That type of malfunctioning of the government redistributive mechanism is at heart due to differing motivations: the interests of the policy makers and their civil servants do not coincide with those of the public.

"Policy formulation properly requires that the realized inadequacies of market outcomes be compared with the potential inadequacies of nonmarket efforts to ameliorate them."[55] This statement by Wolf reflects the accumulating experience with ill-advised or ill-executed governmental attempts to improve upon the allegedly inadequate functioning of sectoral markets; it also echoes a substantial amount of theoretical and empirical work documenting the logic of nonmarket failure.

The theory of nonmarket failure observes that it is frequently impossible to organize nonmarket (e.g., government-sponsored) mechanisms which would reconcile the calculations by decision-makers of their private and organizational costs and benefits with total (economy- or sector-wide) costs and benefits. Perhaps the most widespread sort of nonmarket failure is the divergence between the officially set objectives of a public intervention program and the internal goals of the bureau or of the agency which is designated to carry out the program. Public objectives will typically suffer from ambiguity in definition and difficulty in measurement. In the absence of good direct-performance indicators, the bureau will set its internal objectives to guide and evaluate its own performance and the performance of its personnel. The consequent structure of rewards and penalties results in an internal version of the price system. For a concrete illustration the reader is invited back to Box 3.

To the extent that the maximization of internal goals is not synonymous with the maximization of public objectives, *internalities* will be present and will as predictably skew the results of nonmarket activities

---

55   Charles Wolf, Jr., "A Theory of Non-market Failure," *Journal of Law and Economics*, April 1979, pp. 107-139.

away from a social optimum as externalities may distort output in private markets. As Wolf puts it, the existence of externalities means that some social costs and benefits are not included in the calculus of private decision-makers. The existence of internalities, on the other hand, results in "private" or organizational costs and benefits being included in the calculus of social decision-makers.

A Last Word on Industrial Policy

As Watson has remarked, it is one thing to recommend wise government, another thing entirely to say how to make it operational.[56] The main theme running through this chapter is that while the logical arguments underlying governmental intervention in favour of innovativeness are quite plausible, their empirical foundations are weaker. The true difficulty comes with the implementation of industrial and strategic trade policies: market failure may be converted into non-market failure.

In the final analysis the debates preceding and the actual setting in place of such policies in Canada so far represent an uneven contest. The proposers and implementers of such policies are full-time, highly paid public servants, backed up by consultants and all manner of techno-science lobbies, whose almost sole task and raison d'être is to dream up ways of spending the taxpayer's money. The resisters appear to be a few editorialists, a handful of academics, some think tanks and, occasionally, a business group, such as the Canadian Chamber of Commerce. As is the case with most policies that are of benefit to narrow sectoral interests, the general public cannot be mobilized to defend itself, since the costs of such defence may exceed the worth of non-intervention in any particular case.

While political competition in the United States and in Britain has shifted the onus of proof on industrial policy advocates, in Canada the restraint comes not from a sound, empirically-based argumentation, but from general deficit-induced restrictions.

---

56   Watson, *op. cit.*, p. 207.

Chapter 3

Innovativeness—An Economic Perspective

"Perhaps the greatest obstacle to understanding the role of innovation in economic processes has been the lack of meaningful measures of innovative inputs and outputs."
Acs and Audretsch[1]

Introduction With Definitions

THE FIRST TWO CHAPTERS PRESENTED an overview of the issues to be discussed in this book and their inherent interest. It is now time to be more specific. In order to understand the soundness or otherwise of public policy interventions aiming at innovative performance, we first have to gain a more thorough comprehension of the innovativeness phenomenon itself. Two viewpoints complement each other: the economic and the managerial. The first pays more attention to the context,

---

1   Zoltan J. Acs and David B. Audretsch, "Innovation in Large and Small Firms,"*American Economic Review*, September 1988, p. 678.

the second to the inner workings of innovativeness. This chapter focuses on the economic perspective.

It is not usual to speak of innovativeness. We use this word here to encompass both the creation and the adoption/diffusion of innovation. As was already stated in chapter 1, innovation is a process which, through technical, industrial and commercial steps, leads either to the marketing of new and improved products or to the commercial use of new and improved production processes, or both. (The research behind such innovations is sometimes called product-oriented or process research).

A linear vision of the place of both innovation and adoption/diffusion is offered in Figure 7. Basic research (or fundamental research), applied research, and development all feed successively into innovation. They are defined as follows:

> *Basic research* includes research projects which represent original investigation for the advancement of scientific knowledge and which do not have specific commercial objectives, although they may be in fields of present or potential interest to the reporting institution.
>
> *Applied research* encompasses research projects which represent investigation directed to the discovery of new knowledge and which have specific commercial objectives with respect to either products or processes.
>
> *Development* is a technical activity concerned with non-routine problems encountered in translating research findings or other general scientific knowledge into products or processes.[2]

There are two points to note in connection with the definitions. The first is the difference between basic and applied research, a distinction that hinges on the researchers' motivation: new knowledge for its own sake, or practical objectives. The second is that development does not include marketing research or other work concerned with market development.

---

2   These are definitions adopted by the OECD Frascati agreements.

Figure 7: A Linear View of Innovation/Adoption

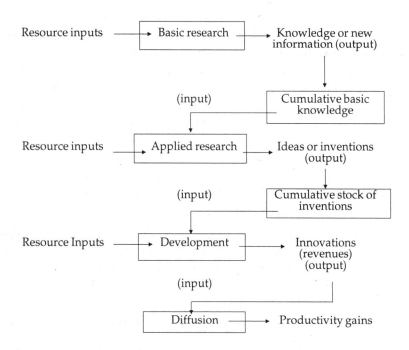

Based on Ryuzo Sato and Gilbert S. Suzawa, *Research and Productivity*, Boston: Auburn House, 1983, p. 49.

An important implication of these definitions is the distinction between invention, research and development, and innovation. Invention, whether representing a major or a modest advance in industrially useful knowledge, is merely an idea or a model which must be further developed, most often with the assistance of research and development activity, to reach a stage of technical feasibility. Once there, products and processes earn their adjective "innovative" only when they have been launched, with or without final commercial success, into a firm's market or into the firm's own production system. Furthermore, innovation is deemed to generate revenues for its creator (product innovation) and decrease costs, or equivalently yield productivity gains, for the customer-adopters to whom it has diffused. While this representation of the innovative process may be pedagogically advantageous, it is certainly oversimplified.

Thus, Mowery and Rosenberg remind us that Sadi Carnot, a French engineer, while working on improving the efficiency of steam engines, created the science of thermodynamics.[3] Half a century later, in 1870, Pasteur was trying to solve practical problems with fermentation and putrefaction in the French wine industry. Along the way he invented the modern science of bacteriology. In the 1930's, work at Bell Labs on static in radiotelephone transmission resulted in the birth of radioastronomy. In general in industries at the edge of technology the problems encountered may be so new as to require a general rethinking of existing principles, or basic scientific research. Often, then, empirical technological challenges of the commercial kind lead to fundamental reflection, or to fundamental research.

Mowery and Rosenberg also point out that the distinction between basic and applied research turning on motivation is not robust. Individual scientists in private industry may genuinely pursue the advancement of knowledge per se, while their managers may be strongly motivated by expectations of useful findings.

It is thus possible that there is a very blurred line between research results that are appropriable and those that are not, despite the general belief that fundamental research falls into the latter category. As pointed out in the preceding chapter, the notion of non-appropriability of research results may be further weakened by the realization that basic and applied research may be required to enable recipients of potential spillovers to absorb such.

While the distinctions between fundamental and applied research are not what they are cracked up to be, their basic employment and usefulness cannot yet be discarded.

---

3   David C. Mowery and Nathan Rosenberg, *Technology and the Pursuit of Economic Growth*, Cambridge: Cambridge University Press, 1989, pp. 12-13.

Table 5: Expenditures on the Development of Innovations by Process Stage for 234 Innovations*

Stage	169 Products		65 Processes		All Innovations
	$ Million[a]	%	$Million	%	%
1. Basic Research	5.2	4.0	13.6	2.7	3.0
2. Applied Research	13.6	10.3	29.2	5.9	6.8
3. Development[b]	52.4	39.8	150.4	30.1	32.1
4. Manufacturing Start-up[c]	53.6	40.8	294.4	58.9	55.1
5. Marketing Start-up	6.8	5.2	11.9	2.4	3.0
Total	131.6	100.0	499.5	100.0	100.0

* The industries: telecommunications equipment and components, electrical industrial equipment, plastic compounds and synthetic resins, non-ferrous smelting and refining, crude petroleum production.
[a] Current dollars—for innovations launched anywhere between 1961 and 1979.
[b] Includes engineering, layout, design, prototype construction, pilot plant construction, testing, etc.
[c] Includes tooling, plant arrangement, construction of additional plant, acquisition of equipment etc.

Source: Dennis P. De Melto, Kathryn E. McMullen and Russel M. Wills, *Preliminary Report: Innovation and Technological Change in Five Canadian Industries*, Economic Council of Canada Discussion Paper No. 176, Ottawa: October 1980.

The Economic Council of Canada and the federal Department of Industry, Trade and Commerce carried out a survey of profitable product and process innovations in five Canadian manufacturing industries in 1980.[4] The information, gathered about some 230-odd innovations, provides us with reasonable estimates on the relative outlays necessary to launch a new product or process—as outlined in Table 5.

The figures in Table 5 are largely in conformity with North American patterns of industrial-goods manufacturers. (For a detailed anatomy of innovation expenditures consult Mansfield et al.[5]) Between 40 and 50 percent of the total launch cost is accounted for by R&D expenditures, with most of the remainder consisting of manufacturing start-up costs. (We would expect marketing start-up costs to be much higher in consumer-goods industries.) Manufacturing start-up-takes a larger percentage in process innovations—the result of research aimed largely at within-firm cost reduction—than in product innovations which stem mostly from sales revenue-oriented research.

As expected, basic research activities that could be assigned to the reported innovations represented but a small fraction (3 percent) of the total investment; even as a fraction of total R&D they amounted to only 7 percent. Industrial firms tend to spend relatively little of their research funds on basic research, proportionally more on applied research, and the most on development.

Five years later, Statistics Canada reported a more generous engagement in fundamental research, still in the private sector, but this time by industry (see Table 6). While the first table shows more money spent on process research, the second implies heavier expenditures on product research. (In dollar terms, not shown in table 6, transportation equipment, electronics and chemicals dominate in R&D outlays.)

A third table (7) gives yet another aspect of the part that fundamental research plays in the overall research effort—this time of nations (if, of course, it can be taken for granted that government does finance most of the basic research going on).

---

4   Dennis P. De Melto, Kathryn E. McMullen and Russel M. Wills, *Preliminary Report: Innovation and Technological Change in Five Canadian Industries*, Economic Council of Canada Discussion Paper No. 176, Ottawa: October 1980.

5   Edwin Mansfield et al., *The Production and Application of New Industrial Technology*, New York: W.W. Norton, 1977.

Table 6: Current R&D Expenditures, by Application and by Industry, 1986

Industry	Basic Research	Products	Pro-cesses	Tech-nical Services
	% of budgets			
Food, beverages and tobacco	9	61	20	10
Wood, pulp and paper	13	34	43	11
Mines, metals and non-metallic mineral products	3	38	49	10
Transportation equipment	9	87	2	1
Electrical and electronic products	12	83	3	2
Oil wells and petroleum products	7	38	36	20
Chemical products	4	73	17	6
Other manufacturing industries	8	70	20	2
Engineering and scientific services	8	60	7	24
Other service industries	9	51	15	25
Total	9	69	13	9

Source: Statistics Canada, *Science Statistics Bulletin*, Ottawa: February 1988, p. 3.

There is good logic behind the fact that most of the basic R&D goes on in the non-profit sectors. In advanced industrial countries there are several publicly supported sets of institutions devoted to "pure" advancement of knowledge, foremost among them the universities and

Table 7: Non-oriented Research Programs as a Percentage of Civil Government Outlays on R&D

Country	1983	1986	1989
Belgium	21.4	23.8	25.7
Canada	23.0	20.7	22.5
France	23.9	21.8	24.6
Germany	12.3	13.9	16.7
Italy	6.2	7.2	10.9
United Kingdom	13.3	8.3	8.5
United States	10.9	11.5	11.1

Note: France, the U.K. and the U.S. spend up to ⅔ of government- financed R&D on defence-oriented objectives. Much fundamental research is going on under that label.

Source: OECD, *Main Science and Technology Indicators*, Paris, June 1990, p. 69.

government-sponsored research bodies, such as the National Health Institutes in the U.S., the National Research Council in Canada or the Max Planck Institutes in Germany.

Working on projects with barely identifiable commercial pay-offs, they nevertheless make research results actively and willingly available to the industrial sector where "practical" pay-offs may be perceived. This pattern of publicly supported non-profit research organizations devoting themselves to fundamental research has evolved for three reasons. The oldest of those reasons is the emergence of the university as the spearhead of basic advances in human knowledge.

The second and third reasons overlap: The commercial payoff from fundamental research is so uncertain that a profit-oriented enterprise finds it difficult to justify investment in it. Because basic research is perceived as having high social returns (as opposed to private returns accruing to a firm only), society is willing to support it when private business is not. Yet this traditional explanation, one might almost say old-fashioned explanation, should be modified in light of the Mowery-Rosenberg findings discussed on a preceding page.

Despite its aversion to basic research with its vague pay-off possibilities, private industry incurs substantial risk in its innovative activities both with regard to applied and developmental research and with respect to launch outlays. Mansfield gathered data from 16 U.S. industrial firms, among whom there were four drug and two electronics producers, and derived from each firm certain average project completion probabilities based on a number of successful and unsuccessful ventures.[6] He made the distinction between technical completion (the R&D project achieves its technical objectives), commercialization (the new product or process is carried beyond test-market or pilot plant trial) and economic success (a rate of return on the project exceeding other non-R&D alternatives.)

Among the 16 firms, each of whom have carried out a large number of new-product or new-process projects, the average probability of technical completion was a reasonably "safe" 57 percent. Given technical completion, the probability of commercialization was 65 percent. (This means that 0.57 times 0.65, or only 37 percent, of projects launched were commercialized.) The probability of economic success, given commercialization, was reported (by 11 of the 16 firms) to be 74 percent. Thus, on the average among this sample of firms, *only* 27 percent (0.57 times 0.65 times 0.74) of projects undertaken in the lab returned an economic profit. In a previous investigation of 220 projects completed during the 1960s in a chemical and two pharmaceutical laboratories, the odds that a project would return an economic profit on its investment were 12 in 100, or one in eight.[7]

In view of the inherent riskiness of innovative enterprise and of the putative high social—though not necessarily private—returns to it, substantial funding to industry and to non-profit oriented research is

---

6   Edwin Mansfield, *op. cit.*

7   Edwin Mansfield, *Research and Innovation in the Modern Corporation*, New York: W.W. Norton, 1971, ch. 3.

provided from the public purse.[8] Box 4 gives both a numerical and a graphical expression to this statement. Note that the statistics do not include tax alleviations granted to the private sector, on the order of $800 million in 1989.

Technological Innovation

Technological innovation, as we said, comes in two guises. It is either a change in the production function of the innovator or the offer of a new or improved product or service. The first, a process innovation, will decrease the costs of the innovating firm—and possibly enhance its revenues if it licenses the process to non-competing firms. The second, a product innovation, will enhance the innovator's revenues and induce changes in the production function (lower costs) of his customers.

The graphical representation of a process, cost-reducing, productivity-boosting innovation is in Figure 8 which has panels (A) and (B). To understand the figure we must first give some background information.

The explanation of productivity improvement (cost reduction) rests on the notion of the production function. For a given state of technology, such a function shows the maximum rate of output of a product per period that can be attained from given amounts of inputs. Suppose that the product can be manufactured with the help of two inputs, labour and capital, in a process that can be more or less intensive in one or the other factor of production, depending on the factor's relative prices. In other words, capital (K) and labour (L) can be to some extent substituted for each other, if an output (Q) is to be produced in a given period.

Mathematically this can be written as:

(1) $Q = A K^a L^b$

with A as constant. (Note that this multiplicative, so-called Cobb-Douglas production function requires a positive level of each input to give any output). Figure 8, panel (A) shows curves, called isoquants (iso

---

8 While most studies of the failure of new products conclude that the launching of innovative products is risky, there appears to be no study *comparing* risks in other firm endeavours, such as the hiring and turnover of employees.

= same, quant = quantity), which are a locus of all combinations of capital and labour yielding a given level of output in a period.

Let us assume that the production unit's output is 100 units per month and that it is achieved by some combination of capital inputs (such as machine time) and labour inputs, determined by the isoquant

---

Box 4: Expenditures on R&D, by Performing and Funding Sectors, 1991

	Performer							
Funder	Fed-eral	Prov-incial	PRO	BE	Uni-ver-sity	PNP	Total	Distri-bution
($ millions)								(%)
Federal	1515	—	7	465	753	40	2780	29
Provin-cial	—	245	54	68	321	23	712	7
PRO	—	—	2	—	—	—	2	0
BE	—	—	27	3809	205	14	4055	42
Univer-sity	—	—	—	—	1059	—	1059	11
PNP	—	—	—	—	137	59	196	2
Foreign	—	—	5	883	13	9	910	9
Total	1515	245	95	5225	2488	146	9714	100
Share of Total (%)	15	3	1	54	26	1	100	

BE—business enterprise
PNP—private non-profit organization
PRO—provincial research organization

Source: Industry, Science and Technology Canada, *Selected Science and Technology Statistics, 1991*, Ottawa, 1991.

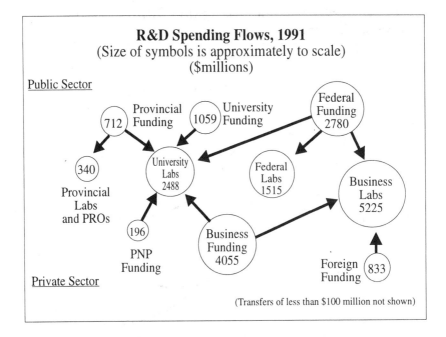

R&D Spending Flows, 1991
(Size of symbols is approximately to scale)
($millions)

Public Sector

Provincial Funding 712 — University Funding 1059 — Federal Funding 2780

340 — Provincial Labs and PROs

University Labs 2488 — Federal Labs 1515 — Business Labs 5225

196 — PNP Funding

Business Funding 4055

Foreign Funding 833

Private Sector

(Transfers of less than $100 million not shown)

curve 100^s, where S signifies "start." Suppose now that the maintenance department discovers a novel way of reducing machine down-time and the same output can be achieved with a lower level of both inputs, though this does not affect the relative productivity of capital and labour. We can now write:

(2) $Q = (A + c) K^a L^b$

and represent this neutral technological change by the isoquant 100^n. The isoquant is southwest of and parallel to 100^s, an indication that the same output as before, 100 units, can now be produced with fewer units of input. This outcome conforms to the definition of increased productivity, which is either a higher output per unit of input, or an unchanged output for a smaller quantity of input. Here the relative productivities of K and L remain unaltered.

Suppose now that, stimulated by continuing wage demands, the production unit is able to purchase and integrate with the existing process additional equipment which increases considerably productive

Figure 8: Productivity-Increasing and Cost-Reducing Technological Innovation

Panel A

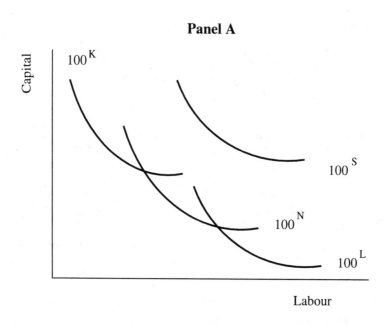

Panel B

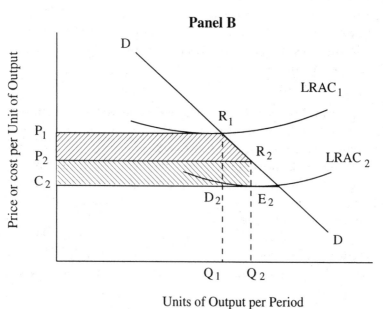

capacity without claiming supplementary staffing. This could be indicated as:

(3) $Q = A K^{(a+c)}L^b$

and is represented by the isoquant 100^k. Capital (equipment), more productive after innovation, results in labour-saving and, because more of K units are likely to be employed now, the innovation can be designated as capital-using (or labour-saving). A similar reasoning stands behind a labour productivity-increasing innovation (100^L).

Since in the wake of the innovation process more units of the (existing) product can be manufactured, or fewer inputs for an existing level of output can be produced, the unit costs of production will of necessity decrease. This phenomenon and its consequences are depicted in Figure 8, panel (B).

Panel (B) indicates some of these phenomena. Before the innovation was in place, a unit of the innovator's product sold at a price of P_1, which covered the long-run average costs (variable, fixed and normal profit), $LRAC_1$, at the output Q_1. The innovation lowers the costs of production as indicated by the downward shift to $LRAC_2$. The innovator, instead of satisfying demand of Q_1 units at price P_1—and reaping profits indicated by the area $P_1R_1D_2C_2$—reduces price to P_2. This induces customers to expand their purchases to Q_2 units per period. The innovator is satisfied with profits $P_2R_2E_2C_2$. But to his private economic profit, indicated by the corresponding area, are joined benefits accruing to his customers. These are depicted by the area $P_1R_1R_2P_2$, an increase in what is called "consumer surplus": customers are now buying (Q_2-Q_1) more units per period and for each they pay (P_1-P_2) less.

We shall come back to panel (B) when discussing social returns. Now let us turn attention to the revenue-increasing innovation. The firm is at first a member of a perfectly competitive industry and earns no *economic* profit, as depicted in panel (A) of Figure 9. At the prevailing market price the firm just covers its marginal and average cost. We could imagine the product to be a run-of-the mill antibiotic.

Let the firm now offer a new antibiotic, protected by a strong patent (see panel B). For a number of years it will be a monopolist, a sole seller in the new drug market. It equates its marginal cost (MC) with marginal revenue (MR) (different now from demand, D, or average revenue, AR)

Figure 9: A Firm in Competitive Equilibrium and After Obtaining a Patent

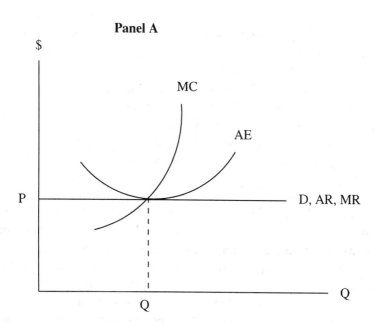

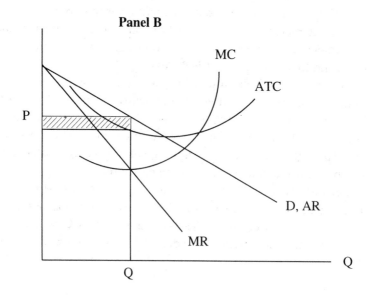

and earns above-average (monopoly) profits, indicated by the shaded rectangle that delineates price less cost times units sold.

Those monopoly profits are presumably the stimulus for past and future innovative activities. On the other hand, the unit price is higher than it would be under competition (where price = marginal cost) and fewer units are as a consequence purchased by the market. This profit-price trade-off represents an eternal bone of contention between consumer advocates and would-be innovation stimulators. A big fight between the two camps erupted on the Canadian scene in the mid-80s, as will be discussed in chapter 6.

The Adoption and Diffusion of Innovations

The speed with which new products or processes, embodied in innovative equipment or intermediate inputs, are adopted by customer industries or households has a major impact on productivity growth in the economy. Similarly, slow or fast diffusion of consumer product innovations clearly makes an important difference to the enhancement of consumption quality. Indeed, there is a Canadian viewpoint which holds that the characterization of Canada's technological performance as a failure in innovation might be as readily identified with failure in the adoption of technology.[9]

The differences in the diffusion speed of an innovation are easily discernible in Figure 10, taken from a celebrated study of hybrid corn by Griliches.[10] It is evident that Iowa farmers were much quicker to adopt the superior corn than those in Alabama. Much of the difference between the rates of adoption among the farming areas could be explained statistically by demand and supply factors.

Thus the original starting date of the S-shaped curve for Wisconsin, about 1932, indicates that hybrid seed companies bred and offered the

---

9   Steven Globerman, "Canadian Science Policy and Technological Sovereignty," *Canadian Public Policy*, Winter 1978, pp. 34-45.

10   Zvi Griliches, "Hybrid Corn: An Exploration in the Economics of Technological Change," *Econometrica*, October 1957.

Figure 10: The Diffusion of an Innovation: Percentage of all Corn Acreage Planted to Hybrid Seed

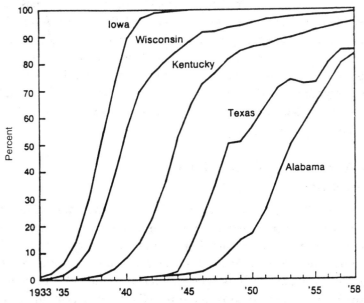

Source: Zvi Griliches, "Hybrid Corn: An Exploration in the Economics of Technological Change," *Econometrica*, October, 1957.

new strains to farmers in that state much earlier than in Texas (1941). The suppliers, who had to breed hybrids to area specifications, chose to start with the heart of the cornbelt first.

The speed of adoption and the proportion of total acreage planted correspond to the slope and the ceiling of the curve. Wisconsin farmers, planting a higher percentage of their land to corn faster than Texans, perceived a greater economic advantage to experimenting with hybrids, since even a minor improvement in yield resulted in substantial returns.

Mansfield, one of the keenest analysts of the diffusion process, proposes four principal factors that govern the speed with which the innovation's level of utilization approaches market saturation: (1) the extent to which the innovation offers an economic advantage over existing products or processes, (2) the extent of uncertainty surrounding early use of the innovation, (3) the rate of reduction of the initial uncertainty regarding the innovation's performance, and (4) the size of the investment required to try out the innovation.[11] Two other factors, among others, are also noteworthy influences in the diffusion process: the extent of the seller-innovator's salesforce and advertising effort, and the degree of competitive pressures to which the customer industries are subjected.

We have good insights into the adoption and diffusion of innovative industrial processes, such as continuous annealing, high-speed bottling, and others, and some understanding of the rate of imitation by competitors of new industrial products. Given this knowledge it is possible to make rough guesses of the economic gains brought about by an accelerated rhythm of an innovation's adoption among industrial customers. However, to the extent that diffusion is, like the earlier stage of innovation creation, a learning process that takes place among users, it can be assisted by a more efficient provision of information. This may be a legitimate goal of government support by means of demonstration centres, early experimental purchases, and other means by which to play upon the four factors mentioned above.

As regards new consumer products, the adoption process and its determinants have been extensively investigated. What is perhaps less well known are the costs imposed by regulation which actually retard the diffusion process. The most glaring instance is that of new drugs widely used in the sophisticated economies of Western Europe whose tardy arrival—by five to ten years—on North American shores results in unnecessary suffering and economic loss.[12]

Another point with regard to innovation diffusion merits attention in the Canadian context. It is generally acknowledged that one of the great potential benefits of the presence of multinational corporations is their technology transfer activity to host country subsidiaries.[13] Even

---

11   Edwin Mansfield, *The Economics of Technological Change*, New York: W.W. Norton, 1968.

12   Sam Peltzman, "The Diffusion of Pharmaceutical Information," in Robert B. Helms (ed.), *Drug Development and Marketing*, Washington, D.C.: American Enterprise Institute, 1975.

13   Harry G. Johnson, *Technology and Economic Interdependence*, New York: St. Martin's Press, 1975, ch. 5.

though subsidiaries may be restricted in their export activities, they often are not charged a full fee for technology transfer from headquarters, enjoying easy and less costly access to the benefits of their parents' research. Therefore, it would seem that any policies directed against a foreign industrial presence will retard the innovation diffusion process.

Diffusion of technology also appears to play a very important part in the overall economic growth of countries. Intuitively we all feel that Japan's spectacular growth was due to its fast imitation (= adoption) of U.S. technology, at least in the fifth and six decade of this century. Economists studying growth speak of technological "catch-up," or of closing the technology "gap." The basic idea is that provided a country has some technological absorption capacity and is willing to foster it, the adoption of foreign technology will speed up productivity increases and so economic growth.

Fagerberg—see Box 5—constructed and estimated technological gap models using data from 1960 to 1983 from 25 industrial countries.[14] His estimates indicate that the more distant a country is from the technological leaders, the faster it tends to catch up. As it catches up, its progress slows down: it becomes more costly to scale technological—and so presumably economic—heights. An indirect confirmation of this assertion is the influential article of Cohen and Levinthal, mentioned in chapter 2 in connection with spillovers.[15] These authors stress—and confirm by their findings at the microeconomic, individual line-of-business level—that the costs of knowledge transfer can be substantial and in part determined by the ease with which learning may occur. Clearly, the more complex the technology is, the less easy it is to master.

---

14 Jan Fagerberg, "A Technology Gap Approach to Why Growth Rates Differ," *Research Policy*, 16, 1987, 87-99. See also, for an indirect confirmation, John W. Kendrick, "International Differences in Productivity Advance," *Managerial and Decision Economics*, Special Issue, 1989, pp. 13-18.

15 Wesley M. Cohen and David A. Levinthal, "Innovation and Learning: The Two Faces of R&D," *Economic Journal*, 99, September 1989, pp. 569-96.

Box 5: A Technology Gap Model Estimated

Hypothesis

A country facing a technological gap (being on a lower technological level than leading innovative countries) can increase its rate of economic growth through imitation (by "catching up").

Econometric tests

Sample is 25 industrialized and industrializing countries with data averaged 1960-68, 1969-73

$$GDP = 5.78* - 0.31TG* + 0.13PAT* + 0.14INV* - 0.09W$$

$$R^2 = 0.60 \text{ N} = 49$$

* means significant at least at 5%

Same sample with data averaged 1974-79, 1980-83

$$GDP = 0.32 - 0.11TG* + 0.11PAT* + 0.22INV* + 0.59W*$$

$$R^2 = 0.70, \text{ N} = 50$$

Variables

GDP_i = growth of GDP in country i in constant prices

TG_i = GDP per capita in i in constant 1980 market prices

PAT_i = growth of patent applications from residents of i
     in other countries

INV_i = gross fixed investment in i as % of GDP in constant prices

W = growth of world trade in constant prices

All variables, with the exception of GDP and world demand are expressed as the difference between the value of the variable for country i and the average value of the variable for all countries in the sample.

Interpretation

The technology gap model "explains" a large part ($R^2 = 0.6$ or 0.7) of the actual differences in growth rates, both between countries and between periods.

The scope for imitation is proxied by TG, the per capita GDP, which in turn proxies the productivity level in country i: the coefficients are -0.31 in 1960-73, -0.11 in 1974-83.

As per capita GDP rises, imitation or adoption of advanced technology becomes more difficult or costly; the catch-up slows down (the early period coefficient, -0.31, is three times as large as the -0.11 coefficient in the later period).

At the same time, patenting (a proxy for own innovativeness) and investment (a proxy for the "embodied" adoption of technology in the guise of new equipment) also contribute to GDP growth. World trade growth contributed significantly to GDP growth in the second period.

Source: Jan Fagerberg, "A Technology Gap Approach to Why Growth Rates Differ," *Research Policy*, 16, 1987, pp. 87-89.

Some Effects of Innovation and Its Adoption

Obviously the primary and most important effect is the innovation's impact on the firm's bottom line, as achieved either by cost decrease or revenue increase. Next come the effects of its adoption on customers or imitators, measured by their productivity change or by "social returns." Finally, the effects diffuse throughout the economy and contribute to the welfare of other nations as well.

Innovation Creation

Let us start rather briefly with the profit impact (more will be said on this subject in the next "managerial" chapter). Sometimes it is possible to evaluate profit impact directly, when the innovative product or process presents a breakthrough and grows fast to become a major part of the corporation's activities. Smith Kline and French's ulcer-curing drug "Tagamet," and Michelin's radial tire are examples.

In most instances, however, the improvements are of a minor, steady nature and profit changes are difficult to trace back to them. Perhaps the most obvious way to circumvent such obstacles is to take the firm's R&D expenditures as a proxy for innovativeness, capitalize them and then compare the rate of return (ROI) on this intangible asset to the ROI on conventional, tangible assets.

Given, however, that ROI does not tell us anything about the firm's R&D variability or the degree of risk that it is subjected to, a more comprehensive approach is to take account of the stock market's evaluation of a company's R&D success. Englander, Evenson and Hanazaki examined stock market data from a very large number of "high-tech" (research-intensive) firms from 1970 to 1986:

> To the extent that new technology translates into higher profits for innovating firms, market analysts have an incentive to identify these firms, and the expected future profits will be discounted into the current price of the firms' shares, even though current sales and earnings are low. Hence, the price/earnings (p/e) ratios of high-tech firms may serve as an indicator of how

much new technology the market expects these high-tech firms to produce in the future.[16]

Over this period the p/e ratios varied from 1.1 to 1.5 times the overall market index. When New York Stock Exchange-listed high-tech firms were excluded, the p/e ratios of the more "junior" high-tech stock varied from 1.2 (1976) to 7.4 (1986) times the overall market index.

In Canada, Johnson and Pazderka estimated the impact of R&D intensity (R&D spending as a percentage of book value) on the (stock market-determined) market value of a sample of about 50 firms.[17] Statistically significant results indicate a positive, fairly large influence exerted by research intensity on the valuation of the firms by the stock market investors over the period 1985-1988.

Can we therefore conclude that innovation-creation is invariably, or at least as a rule, a royal road to business success? Not entirely. There were no failed firms in the samples examined. And—as far as this writer can tell—there is no record of scholarly examination of the riskiness of R&D enterprise to the outside investor. More on all this in the next chapter.

Figure 8, panel (B) showed how a cost reduction consequent upon a process innovation benefits both the innovator—by enabling him to turn an economic profit—and the customers, by offering them a lower price on the existing product. We shall dwell on the reason why the innovator did not appropriate to himself as profit the area $P_1R_1E_2C_2$, but only $P_2R_2E_2C_2$ in a later chapter. Here it is merely to be pointed out (again) that customers received a "consumer surplus" corresponding to the area $P_1R_1R_2P_2$. Except for the last one of them, each one would have been willing to pay a higher price than P_2 per unit; the addition of all

---

16   A. Steven Englander, Robert Evenson and Masaharu Hanazaki, *R&D, Innovation and the Total Factor Productivity Slowdown*, Paris: OECD Economic Studies No. 11, Autumn 1988.

17   Lewis D. Johnson and Bohumir Pazderka, "Firm Value and Investment in R&D," in *Managerial and Decision Economics*, v. 14, 1993, pp. 15-24. The dependent variable is defined as (average) price per share times the number of shares outstanding.

the sales corresponding to the higher prices—up to P_1—represents in value what is called consumer surplus.

Innovation Adoption/Diffusion

We have just pointed out the benefits of cost-reducing innovations to downstream customers. The bulk of economic improvements originating in technology stems undoubtedly from the purchase of innovative products. By definition innovative products, unlike the existing products sold at a lower price in the wake of cost-reducing innovations, require that the customer modify his existing production processes. This is why the word adoption, implying an active participation, is used.

The bank may need to train its staff to use a new software program thereby yielding a richer customer-base information bank. The farmer may follow new planting instructions for the better seed. The consumer must learn how to operate his microwave oven or compact disk player. In all instances the factor input mix in the production process is modified and in all instances, when the innovative product performs as expected, overall productivity of the adoption unit increases. This process is the same as that described in panel (A) of Figure 8.

The productivity increase in customers' operations is well documented for the commercial firm, hardly at all for the private household. The leading savant here is Terleckyj.[18] He examined, in a production function context, the influence of own-performed as well as supplier-performed innovative activity. Some of his results were already presented in Box 2. Here a more detailed explanation is offered.

More precisely, Terleckyj analyzed data from 20 U.S. manufacturing industries. An estimate was made of the (regression) relationship between the average rate of growth in productivity during the years 1948-1966 and 1958 R&D intensity (research cost to value-added ratio) in a highly ingenious manner.

---

18   Nestor E. Terleckyj, "Direct and Indirect Effects of Industrial Research and Development on the Productivity Growth of Industries," in John W. Kendrick and Beatrice N. Vaccara, *New Developments in Productivity Measurement and Analysis*, Chicago: University of Chicago Press, 1980.

Terleckyj used as R&D inputs expenditure (ratios) incurred both within the industry, and by that industry's suppliers in proportion to the share of the vendor industry sales going to the buyer industry. In this way the frequent and valid observation that an industry's productivity depends on technology embodied in purchased capital equipment and supplies in addition to its own technology upgrading efforts was accommodated. In a second round of estimates Terleckyj divided "direct" and "indirect" research expenditures into government-funded and industry-funded categories. His results indicated that returns to private direct R&D were about 30 percent in terms of productivity growth in these twenty industries. Indirect returns, that is returns to "purchased" R&D, were almost three times as high. No statistical indication of significant returns to government-funded direct or indirect research was found.

It is likely that these results on the whole underestimate the effect of R&D since, as repeatedly stressed, the production function approach neglects product quality changes. It is also possible that returns to government funding were underestimated if government support stimulated private R&D expenditures that were complementary to government-sponsored research. It remains that R&D's impact on productivity in U.S. manufacturing was impressive.

Economic Council of Canada researchers, Postner and Wesa, inspired by Terleckyj's methodology, tried to throw light on the relationship between the growth rates of productivity and industrial R&D in Canada.[19] (Please see Box 2). The study is so complex and extensive that its summary is taken verbatim from the Economic Council's umbrella consensus paper *The Bottom Line*:

> ...was done for 13 manufacturing industries over the period 1966 to 1976, with both internal and contracted-out R&D being measured for each industry. The input-output method of measuring productivity growth was used in combination with the technique of multiple-regression analysis. The latter enabled us to distinguish the effects of this industrial R&D from the other variables known to influence productivity growth—i.e.,

---

19   H. Postner and L. Wesa, *Canadian Productivity Growth: An Alternative (Input-Output) Analysis*, Economic Council of Canada, Ottawa, November 1983.

changes in the amount of capital equipment of a given type and changes in the scale of operation. The results showed that R&D done in any given industry had little effect on its productivity growth but had a favourable influence on productivity growth in the industries that are supplied directly or indirectly by that industry. This means, essentially, that most of the industrial R&D in Canada that has a productivity-raising impact is oriented towards creating improved equipment and products for sale to other firms, rather than the development of new production processes that will be used internally by the industry performing the R&D.[20]

Innovation and Trade

One does not need to invoke neofactor proportions or neotechnology theories of international trade to sense that innovative products or lower-cost existing products give firms and industries an advantage in foreign markets.

The connection between trade balance and innovativeness can be represented in several guises. The proxy for innovativeness can be, for instance, R&D expenditure (an input measure) and productivity (an output measure). A vigorous investigation of the influence of R&D effort in a multifactor context on the foreign *trade* of 13 Canadian manufacturing industries was conducted by Hanel.[21] One of the most enlightening aspects of his study concerns Canadian export performance in its largest market, the United States. Hanel used as an index of that performance the share of United States imports in each of the 13 product groupings that Canadian manufacturing held against seven major industrial rivals, namely Japan, Belgium, Germany, France, Italy,

---

20 Economic Council of Canada, *The Bottom Line: Technology, Trade and Income Growth*, Ottawa, 1983, 27.

21 Petr Hanel, *The Relationship Existing Between the R&D Activity of Canadian Manufacturing Industries and Their Performance in the International Market*, Research Report, Dept. of Industry, Trade and Commerce, Ottawa: August 1976. The industries: food and beverage, textiles, clothing, wood products, paper, petroleum products, chemicals, rubber, non-metallic minerals, primary metals, metal products, machinery, electrical equipment, transport equipment except aircraft.

the U.K. and Sweden, in 1969. He then related Canada's shares of the U.S. import markets to four measures of international competitiveness suggested by trade theory, read as of 1976:

Share of U.S. imports held by Canadian manufacturing sector *i* is a function of:

1. Relative sectoral price level proxied by

   (ALGEBRAIC SIGN -, STATISTICALLY SIGNIFICANT)

   WAGES PER UNIT OF OUTPUT IN CANADA/SAME IN U.S.

2. Relative sectoral labour productivity proxied by

   (SIGN +, NOT SIGNIF)

   VALUE ADDED PER EMPLOYEE IN CANADA/SAME IN U.S.

3. Relative sectoral tariff protection proxied by

   (SIGN +, SIGNIF)

   NOMINAL TARIFF RATE IN CANADA/SAME IN U.S.

4. Relative sectoral R&D intensity proxied by

   (SIGN +, SIGNIF)

   R&D INTRAMURAL EXPENDITURES IN CANADA/SAME IN U.S.

(A fifth variable, proxying U.S. control of the Canadian sector, was also included; SIGN +, NOT SIGNIF).

Given that over 90 percent of the fluctuations in U.S. import shares was "explained" (R^2 = .92) and that the effect of the R&D factor was highly significant, we can be reasonably satisfied that there is an association between R&D intensity and trade performance.

Interestingly labour productivity, a partial stand-in for "received" technological innovation, does not show up as statistically significant. Association does not, however, necessarily indicate causation. It is perhaps more prudent to envisage a scenario in which an original human or resource endowment gave rise to an industry successful in exporting part of its output and keeping its internationally competitive edge by investing in research leading to innovation.

While innovativeness in the above example was approximated by its input determinant, R&D, Soete used patents—an output approximation of innovation—to explain international trade patterns.[22] For each of the 40 manufacturing industries in 22 OECD countries he regressed the share of each country i exports of industry j in total OECD exports of industry j on the share of each country's i 1963-77 U.S.-registered patents in industry j in total OECD 1963-77 U.S.-registered patents in industry j; and on other variables. In other words, for a given industry j in 22 countries i

Export share $_{ij}$ = f (Patent share $_{ij}$, Other variables)

In 28 industries out of the 40, the coefficient of patent share proved to be statistically significant and of the expected sign, thus confirming the notion that technological innovation as expressed in patenting confers an internationally competitive advantage. Contrast this, however, with our discussion of the trade performance of hi-tech industries further on in this chapter.

In concluding this section we remind ourselves that both R&D and patents taken out pertain to the "originating" industry and are far from measuring technology's impact in total. We give no estimate here of the adoption or diffusion impact on international trade performance among "receiving" domestic industries. (Though, as indicated by the example in Box 5, there are studies of the impact on industries abroad in a different context).

This brings us to consider again, as already done in part in chapter 1, important measurement issues connected with innovation.

Measurement Issues

We have bumped into several problems already with the measurement—and therefore with appraisal—difficulties with regard to innovativeness. Obviously the most serious one was the fact that innovation, and *a fortiori* innovativeness, can in most instances only be specified

---

22   Luc Soete, "The Impact of Technological Innovation on International Trade Patterns: The Evidence Reconsidered," *Research Policy*, 16, 1987, pp. 101-130.

indirectly, such as by input (R&D expenditure or intensity) or by outputs, such as patents, trade balance etc.

Perhaps the next most difficult aspect is in the evaluation of the effects of innovation if these effects are deemed to be productivity improvements. We have presented, in Table 1, an estimate of the large contribution of technological change to the growth of the national income of the United States. Zvi Griliches, a leading U.S. econometrician and expert in the innovation field believes nevertheless that studies of productivity change seriously underestimate the contribution of innovativeness to economic growth, and that for two reasons.[23] One is the difficulty of estimating correctly the magnitude of spillovers from one firm, industry, or country to another, a topic already touched upon in chapter 2. The other source of difficulty lies in the inability of researchers to measure *productivity* itself correctly:

> The national income accounts, as currently constructed, do not reflect major components of the "product" of R&D and science and hence cannot serve as adequate measures of it. Most of the industrial R&D in the United States has been and is being spent on defense and space exploration connected projects. Its product is "sold" to the public sector and is measured, by accounting convention, by its costs. This implies a zero contribution to measured productivity growth except, perhaps, for its spillover effects on other products and industries.

The third aspect of measurement difficulty is also in part a reflection of definitional problems. Here we refer to the widely bandied-around and perniciously flexible expression "high technology" or simply "high-tech" firms or industries. At first, it would seem that the term is self-defined, that "high tech" designates an entity that relies for its economic success on technology. And this we believe would be the correct view. For instance, nuclear generation of electricity, engineering consulting, stockbrokerages and market research firms all rely on the latest in computer and telecommunication technology to a crucial degree. However, they are rarely, if ever, classified as "high-tech."

---

23   Zvi Griliches, "Productivity Puzzles and R&D: Another Nonexplanation," *Journal of Economic Perspectives*, 2, No. 4, Fall 1988, pp. 9-21.

Hi-Tech

The last three sectors mentioned share the opprobrium of being classified as services and the last two of not undertaking much R&D as
officially defined. It so happens that politicians still suffer from a manufacturing fetish. The result is a wealth of information on that sector,
which accounts for about 20 percent of the gross domestic product, and
very little for services that make up at least three times this proportion.[24]

And so there is little interest in exploring hi-tech aspects in the
service sector. More importantly, however, the sobriquet "hi-tech" is
usually given only to industries whose firms are "own-performed"-re-
search intensive.

Consider the scheme below which shows the possible sources of
new technology for a given firm or industry:
- R&D performed intramurally or funded by industry
- "Invisible R&D imports" from affiliates abroad
- Technology-related payments (royalties etc.) to non-residents
- Government intramural R&D assignable to industry
- University research assignable to industry
- Commercial suppliers' R&D (first upstream level)

R&D performed by the firm or industry, as a percentage of sales or
value added, or R&D personnel as a percentage of employees, is the
standard yardstick by which industries are classified into high, medium, and low-tech. But there is a proliferation of these yardsticks, as
indicated by the 8 types of classification listed in Table 8. All of them are
or have been employed.

An essential point emerges from this rich confusion of definitions
for the Canadian case: the different definitions—and so industry classifications—yield substantially different estimates of export-import trade
balances of high-technology commodities. This is an indicator considered to be very significant by Canadian industrial policy advocates.
(Recall Figure 6).

Now consider the next line in the enumeration. "Invisible" R&D
imports are transfers of technology between multinational affiliates that

---

24 Herbert Grubel and Michael Walker, *Service Industry Growth*, Vancouver:
The Fraser Institute, 1989.

are not formally billed to the receiving units. As will be shown in another chapter, they represent a crucial input into the innovativeness of the heavily foreign-owned Canadian industry. *Neither this item, nor all of the following ones are counted* in the R&D intensity measures used to define "tech" categories.

Technology-related payments do not necessarily signal technological underdevelopment. At least until recently, the Japanese had a large negative trade balance here, while the British had a positive one.

It is suitable now to return to the role of hi-tech industrial groups in foreign trade. Figure 6 showed the negative trade balance Canada has in hi-tech products. But hi-tech products do not an entire merchandise trade balance make. This was shown in a most convincing manner by Hughes, who analyzed the trade patterns of six countries in the 1980s.[25] The three countries with the best *overall* trade performance during this time—Japan, Germany and Italy—had their best performances respectively in high, medium and low technology. The two countries with the worst performance—the U.S.A. and the U.K.—had their best performance in high technology. Hughes concludes that trade performance cannot be explained by reference to specialization by technology group alone.

Elsewhere we documented the importance of research carried out by (federal) government labs and by universities which directly benefit industry, and attempted with some success to assign it to specific sectors.[26]

We have already shown the role suppliers play in selling innovations downstream. We list a separate line for the first upstream level to signal that a powerful connection exists between such suppliers and customers. It is now well known that immediate customers often have a very active hand in the supplier's development of new materials, processes or capital equipment.[27]

---

25   Kirsty S. Hughes, "Technology and International Competitiveness," *International Review of Applied Economics*, 1992, 166-183.

26   Kristian S. Palda, "Technological Intensity: Concept and Measurement," *Research Policy*, 15, No. 4, August 1986, pp. 187-198.

27   Eric von Hippel, "Successful Industrial Products from Customer Ideas," *Journal of Marketing*, January 1978, pp. 39-49.

Table 8: Methods of Classification of High Technology Products used in the United States and Canada and by the OECD According to Conklin and St. Hilaire[a]

DOC1—Developed by Boretsky: This is an SIC/industry-based definition which uses industry R&D expenditures and employment of scientists and engineers to determine which industries are technologically intensive. High technology industries are generally those which usually spend at least 10% of their gross value added product on R&D and/or have at least 10% of their total employment consisting of scientists, engineers, and technicians. ("Concerns About the Present American Position in International Trade," *Technology and International Trade*, National Academy of Sciences, Washington, DC, 1971).

    NSF—Developed by the National Science Foundation: This definition is based on the number of scientists and engineers employed in research and development and company R&D expenditures as a percentage of total sales. R&D-intensive goods are those associated with industries having 25 or more scientists and engineers engaged in R&D per 1000 employees and whose R&D funding amounts to at least 3.5% of net sales. The definition uses industry data and is SIC based. (Science Indicators 1982, National Science Foundation).

    DOC2—Developed by R. Kelly: This definition is based on R&D expenditures relative to shipments. Data used in this definition were product- based. Products having an above average level of R&D intensity were classified as high technology. This definition was originally based on SIC data but an SITC concordance for studying international trade flows was also developed. ("Alternative Measurements of Technology-Intensive Trade," *Staff Economic Report*, U.S. Dept. of Commerce Office of Economic Research, Sept. 1976, and, "The Impact of Technological Innovation on International Trade Patterns," *Staff Economic Report*, U.S. Dept. of Commerce, Office of International Economic Research, Dec. 1977).

[a]Taken from David W. Conklin and France St. Hilaire, *Canadian High-Tech in a New World Economy*, Halifax: Institute for Research on Public Policy, 1988, pp. 136-7.

Table 8 (continued)

DOC3—Developed by Davis: This definition is also based on R&D expenditures as a percentage of shipments but includes not only the R&D spent directly by the final producer, but also the R&D spent by producers of intermediate products. Products having a significantly greater intensity of embodied R&D than other products were classified as high-technology. This definition is SIC based. ("Technology Intensity of U.S. Output and Trade," U.S. Dept. of Commerce, International Trade Administration, May 1982).

OECD1—Products are classified as high-tech on the basis of their R&D intensity, which is computed as OECD-weighted ratios of R&D expenditures and output of each industry. Products are categorized as high, medium, or low-R&D intensity depending on whether this ratio is greater than 4, between 1 and 4, or under 1.

OECD2—This classification is a refined version of OECD1. Only high-tech products are classified. The classification is also based on the R&D intensities of industries but adjustments have been made at the commodity level to exclude specific products of "high-tech" industries which do not qualify as high-tech and to include certain products of "medium-tech" industries which do qualify.

STD1—The classification used until recently by the Science, Technology, and Capital Stock Division was not based on a set of specific criteria to categorize high-tech products. The classification consisted of a relatively broad-ranging list of products considered technology intensive as suggested by the Industry Branch of MOSST in the 1970s.

STD2—The Science, Technology, and Capital Stock Division announced at the beginning of 1987 that it was replacing the STD1 classification with the OECD2 classification. This classification is more up-to-date and more likely to provide comparable estimates across countries. The new list has significantly fewer products categorized as high-tech particularly in the non-electrical machinery and chemical groups.

Source for U.S. Classifications: U.S. Dept. of Commerce, *U.S. High Technology Trade and Competitiveness*. Staff Report, February 1985.

Thus, we conclude that the employment of the "hi-tech" concept is non-operational and perhaps even self-serving when used as a lever to obtain government favours, unless it is the case that "hi-tech" signals so-called strategic industries. Conklin and St. Hilaire cite an OECD 1985 paper that lists the characteristics associated with hi-tech products:[28]

- high dependence on a strong technology base and a vigorous R&D effort
- considerable strategic significance to governments
- long lead-times from basic research to industrial application, short lead-times in commercialization, and accelerated obsolescence under the competitive pressure of new product and process introductions
- high risks and large capital investments
- high degree of international cooperation and competition in R&D production and worldwide marketing

While these characteristics do not constitute tangible criteria, there is a wide consensus, at least among politicians, that some industries are more important than others, that they are the sources of technical progress for the economy, that they are in a sense "strategic." And here opinions have been translated into actual support programs in Canada, France, Japan, and elsewhere, as will be described further on. Some indication of it is contained in table 9.

Richard Nelson is probably the best-known authority in innovation economics which examine the proposition that hi-tech industries are strategic and leading.[29] The proposition that hi-tech industries are leading means that they tend to drive and mold economic progress across a broad front. The idea that hi-tech industries are strategic implies that national economic progress and competitiveness are dependent upon national strength in these industries, and help from government is required to shore up this strength.

---

28   David W. Conklin and France St. Hilaire, *Canadian High-Tech in a New World Economy*, Halifax: Institute for Research on Public Policy, 1988, pp. 136-7.

29   Richard R. Nelson, *High Technology Policies: A Five-Nation Comparison*, p. 1.

Nelson examined only three of the important, conventionally-defined hi-tech industries across the five largest industrial nations: semiconductors and computers, civil aircraft, and nuclear power. He came to the conclusion that the first two were definitely "leading," in that they had widespread economic ripple effects. His opinion on their "strategic" nature was sceptical in that he was uncertain that taxpayer support for them was needed or helpful.

Since in this policy-oriented monograph the question of how useful taxpayer support is to innovativeness is the pre-eminent one, we submit that the concept of high-technology industries is, on the whole, detrimental to policy decision-making.

Table 9: Public Funding for R&D in the High, Medium and Low-intensity Industries as Respective Proportions of Total Public Funding of R&D, 1980
(very approximate estimates)

	High	Medium	Low
United States	88.0	8.0	4.0
Japan	21.0	12.0	67.0
Germany	67.0	23.0	10.0
France	91.0	7.0	2.0
United Kingdom	95.0	3.0	2.0
Italy	69.0	18.0	13.0
Canada	67.0	16.0	17.0
Sweden	71.0	20.0	9.0

Source: OECD, *OECD Science and Technology Indicators No. 2*, Paris: 1986, p. 63.

Chapter 4

Innovativeness—A Managerial Perspective

> *"No government program can substitute for managers, owners, and employees who are not competitive on their own."*
> Terence Corcoran[1]

IN THIS CHAPTER WE LOOK at the firm's management of the generation or adoption of innovative products and processes. Understanding some of the basic managerial issues in this area should sharpen our critical appreciation of the scope and limits of public strategies to stimulate innovation.

Management—The Bridge Between Technology and Markets

It can not be repeated often enough that innovation-oriented research in a firm, or even commissioned by a firm, must be linked consciously and deliberately to commercial considerations. This link is provided by

---

1   *Globe and Mail*, July 22, 1992, p. B2.

management both at the top of the enterprise, and by those in charge of the various functional areas: marketing, production, R&D, human resources, finance, management information systems, etc.

The most recent and probably the best illustration of the performance-tested strong symbiosis between R&D and the other business functions is presented in Box 6. In addition, the diagram shown therein illustrates the mutual influence of "environmental" variables and R&D intensity, in part already signalled in Figure 1.

Apart from the marketing effort-R&D intensity connection indicated in the box, the other highest correlation of R&D is found to be with investment intensity (average investment divided by net sales). This could indicate that R&D efforts are a component of a larger strategic concept in which these efforts are coordinated with other investments. As the authors mention, R&D-intensive businesses must invest in order to capitalize on the opportunities presented by their innovation.

This is clearly another indication of the bridging role that management must assume to make innovation creation or adoption pay off in the market. It is well known that the conversion of a successful R&D investment into a successful venture in the market normally requires a commitment of people and capital many times larger than is required to do the initial R&D. The investment perspective of R&D again allows us to underline the importance of the business climate which either welcomes it or discourages it. It also makes us aware that an investment in innovation must be judged against other investments:

> An expenditure on R&D is an investment, comparable to any other business investment made with the expectation of improving existing products or of developing new products and processes which will increase future profits. The amount to be invested in R&D is determined by comparison with other investment opportunities available to the firm. The attractiveness of R&D is influenced by its characteristics as an investment in comparison to these other opportunities.

> The basic characteristics of R&D investment are the risk, the front-end investment, the delayed return and the time value of money.

> The most important characteristic is that the return from an R&D project often occurs many years after the initial investment. The risk in that investment is higher because there is

Box 6: R&D in Business Strategy

Zif, McCarthy and Israeli set out to investigate the linkage between R&D intensity and management strategy variables in R&D-active business units, holding certain relevant environmental variables constant.[a] Their notion of the interplay of R&D intensity and the two sets of variables (strategic and environmental) is depicted in their Fig. 1, here reproduced.

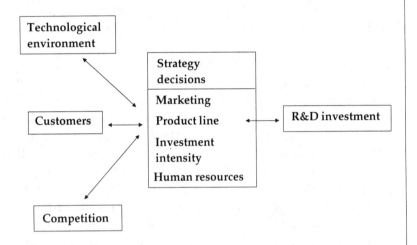

Their data sample consisted of 1,426 industrial strategic business units taken from the 1989 PIMS data bank, with all data averaged over 4 years. (The strategic business units, akin to the Federal Trade Commission's line-of-business units, are components of firms which have their own markets, competitors, and enough leeway to calculate with confidence their profitability. Thus they are more suited to standard economic analysis than the typical modern multiproduct firm).

The empirical relationship between the management decision variables and R&D intensity is set out below in the regression results they obtained (we do not show their regression results with the environmental variables):

[a]J. Zif, D. McCarthy, A. Israeli, "Characteristics of Businesses with High R&D Investment," *Research Policy* 19 (1990), pp. 435-445.

Box 6 (continued)

Multiple regression results with R&D expenditures/ revenues (average of 4 years) as dependent variable

Independent variable	Expected sign	Standardized regression coefficients
Strategy decisions		
Marketing		
1. Marketing expenses/revenue	+	0.194*
2. Export sales	+	0.098*
3. Vertical integration forward	+	0.004
Product Line		
4. Percent new products	+	0.163*
5. Age of product	+	0.101*
Investment intensity		
6. Investment/revenue	+	0.229*
Human Resources		
7. Unionization Rate	−	-0.077*

$R^2 = 0.375$; $F = 65.1$; $N = 1426$.
*$p = 0.05$.

The variables were defined operationally as follows:
- Marketing expenses/revenue—Total marketing expenses divided by net sales.
- Export sales—Total exports of all businesses in the SIC group to area outside the U.S. expressed as a percentage of their total sales both within and outside the U.S.
- Vertical integration forward—Compare the degree of forward vertical integration of this business relative to its three largest competitors (1 = less; 3 = more).
- Percent new products—The percentage of total sales accounted for by products introduced during the three preceding years.

Box 6 (continued)

- Age of product—When were these types of products or services, using the current basic technology, first introduced? (0 = prior to 1930; 7 = 1975+).
- Investment/revenue—Average investment (book value) divided by net sales.
- Unionization rate—The percentage of total employees who are unionized.
- Total R&D/revenue—All expenses incurred to improve the existing products and services or to develop new products or services plus expenses for improving the efficiency of manufacturing and distribution processes, divided by net sales.

Note that the degree of marketing efforts, marketing expenses/revenue, appears to be the second-highest correlate of R&D intensity.

---

uncertainty concerning the financial market and the competitive situation when the results of the R&D program mature. By comparison a production machine has a short commissioning interval, and the risk of significant change in the *business climate* during the procurement and installation time is small.[2]

So a new product or new process is the result of an investment decision on the part of management. It seems natural, therefore, to enquire about the elements of management's deliberations about the desirability of such an investment. The first—and obvious—question that management will ask is whether, given the presumably higher risk levels, industrial innovations are *profitable*.

---

2 Report of the Business Council on National Issues and The Canadian Manufacturers' Association Joint Committee on Industrial R&D in Canada, Toronto: October 1979.

The Profitability of Industrial Innovations

An innovation, it has been said, is a new or modified product or process cast upon the waters of competitive seas. A successful innovation is by definition profitable, as it increases revenues or lowers production costs, but not all or even a majority of innovations are successful and therefore profitable. An innovation typically requires outlays not only for research, but also for "downstream" activities, such as new plant and marketing launch costs. Because the outlays precede the cash flows occasioned by the new product or process by more than one accounting period, we talk about investment in research and other innovative endeavours rather than about (current) expenses. In this sense we may legitimately ask about the profitability of innovation: did the money spent on one or several "lumpy" new-product projects return as much or more than money invested by the firm in joint ventures, new plant, different data processing systems? Because it is very laborious to obtain statistical evidence about returns to individual innovation projects, other means are usually resorted to in order to uncover some "typical" rates of return, and those only to the more frequently accessible data about R&D outlays.

The true profitability of a firm to its shareholders is reflected in the appreciation of the value of stock and the dividends paid out over some considerably long period. In the absence of this ideal measure, profitability is often approximated by some indicator of long-run return on capital employed. Nevertheless, the use of conventional accounting figures to calculate long-run returns carries some well known risks. Apart from the perennial problem of replacement cost, exacerbated by high inflationary rates, the prudent accountant's (and tax collector's) custom of expensing in its entirety investments in intangible assets distorts the picture of profitability of those firms which do research and advertise. In other words, the accumulating—and depreciating—investment in technological knowledge that is industrial research, is not properly capitalized. Nor is the continuing effort of the advertiser to build up the intangible capital of goodwill in the mind of the consumer.

Grabowski and Mueller, integrating a long line of previous studies (including this writer's), have shown a convincing way of calculating long-run rates of return on investment in research and advertising and present evidence that R&D yields above-average profits.[3]

Instead of defining the rate of return (ROI) in the usual way, that is, as:

$$PR_t = \frac{Sales_t - Variable\ Costs_t - d_K K_t - r_t - a_t}{K_t}$$

where

PR_t = "profitability" or ROI in period t,

K_t = total (tangible) assets in t,

d_K = depreciation rate for tangible assets,

r_t = R&D expenditures in t,

a_t = advertising expenditures in t,

Grabowski and Mueller defined profitability as

$$PR_t = \frac{Sales_t - Variable\ Costs_t - d_K K_t - d_R R_t - d_A A_t}{K_t + R_t + A_t}$$

where d_R and d_A are the depreciation rates for research and advertising assets, R and A.

This meant that they had to calculate, for the 86 U.S. corporations in the nine-odd manufacturing industries which constituted their sample, both research and advertising capital and their respective rates of obsolescence or wear out—in short, depreciation. They did so by assuming constant proportional depreciation rates for intangible capital:

$$R_t = r_t + (1 - d_R)\ R_{t-1} \text{ and } A_t = a_t + (1 - d_A)A_{t-1}$$

In this formulation the stock of intangible capital (research or advertising) in a given period t is a sum of the current period's outlay (r_t or a_t) and of the previous period's capital (R_{t-1} or A_{t-1}), suitably depreciated. Using as depreciation rates 10 percent for R&D and 30 percent for advertising capital stocks, Grabowski and Mueller found that the proportion of total assets accounted for by intangible assets reached as high as 30 percent among the seven pharmaceutical firms in the sample. Interestingly, the variance of profitability rates among the

---

3   Henry Grabowski and Dennis Mueller, "Industrial R&D, Intangible Capital Stocks, and Firm Profit Rates," *Bell Journal of Economics*, Autumn 1978, pp. 328-343.

sample firms was reduced by one-half when intangible stocks were capitalized, indicating the degree to which improper accounting treatment of such capital may distort the true economic picture.

Having defined and calculated a more realistic measure of profitability, Grabowski and Mueller related it then to the accumulated intangible assets that the sample firms held in research knowledge and advertising goodwill. In this way they obtained estimates of the long-run rates of return to these two types of investment. Holding some other potential determinants of profitability (market structure, firm size, industry growth) constant, they found an after-tax return on R&D capital of 11.7 percent. This is significantly larger than the average 7.1 percent average after-tax return on total capital employed by the firms in the sample. When the firms were divided into the 39 whose research-to-total-capital ratio was higher than 10 percent, and into others, the research-intensive firms (pharmaceuticals, chemicals, machinery) earned a surprisingly high 16.7 percent after-tax return on R&D spending, while the others (such as paper, metals and petroleum refining) did not obtain higher than average (i.e., 7.1 percent) after-tax returns from research. (Among the companies manufacturing largely industrial (i.e., non-consumer) products, advertising appeared to return above-average returns only in the case of pharmaceuticals.)

While the title of this section specifies *innovation* as the subject, so far only returns to R&D have been dealt with. The reason for this is an absence of statistical studies on innovation's profitability, due mainly to the fact that corporate accounting systems are not set up to report all the (R&D plus manufacturing set-up plus marketing expenditures, etc.) investments connected with putting an innovation on the market. Nor are they usually geared to report on the revenues connected with minor or even major product or process innovations.

Before providing some up-to-date information on R&D profitability, we had better explain the statistically established links between innovations and industrial research. In 1988 Acs and Audretsch examined 4,407 manufacturing innovations recorded by the U.S. Small Business Administration in 1982.[4] In this enormous sample they found a

---

4   The Small Business Administration defines innovation as "a process that begins with an invention, proceeds with the development of the invention,

correlation of 0.746 = r between company-financed R&D expenditures in millions of dollars and the number of innovations launched by these individual firms. (The correlation decreased to 0.481 when company plus government-financed R&D was the measure). Next, assembling the data into 247 four-digit industries, A&A estimated the elasticity of (the number of) innovations with respect to (millions) of company-financed R&D to be about 0.4. These estimates indicate a reasonably strong substitutability of the R&D input measure for the "apples-and-oranges" measure of innovation.

Probably the best overview of recent R&D profitability studies can be found in the not easily accessible *Economie et statistique* journal in an article by Mairesse and Mohnen.[5] They discuss the findings of private rates of return to R&D in 5 studies of *individual firms*, in which the sample sizes range from 5,240 U.S. firms to 135 Japanese firms. The rates of return lie between 11 and 27 percent.[6] When industries rather than individual enterprises are the object of estimation, 4 studies (25 to 193 industries) show that private rates of return to stretch from 12 to 31 percent.

Bernstein, in two articles already discussed in Chapter 2 (see chapter 2, footnotes 17, 19) estimates the average *private* rate of return on R&D in 680 firms in 7 Canadian industries between 1978-81 to be 12 percent, while the average social rate of return (private plus spillover effects) is 22 percent; in 9 Canadian *industries* between 1963-83 these rates are estimated to be 32 and 58 percent.

Finally Griliches,[7] working on a monumental data set of firms ranging from 386 to 652 in size and covering the years 1967 to 1977,

---

and results in the introduction of a new product, process or service to the market." Zoltan J. Acs and David B. Audretsch, "Innovation in Large and Small Firms: An Empirical Analysis," *American Economic Review*, 1988, pp. 675-689.

5   Jacques Mairesse and Pierre Mohnen, "Recherche—Développement et productivité," *Economie et statistique*, No. 237-8, Nov.-Dec. 1990, pp. 99-108.

6   When industry dummy variables are not used in the regressions.

7   Zvi Griliches, "Productivity, R&D, and Basic Research at the Firm Level in the 1970's," *American Economic Review*, March 1986, pp. 141-154.

reached the opinion—similar to the earlier conclusions of Mansfield[8]—that "firms that spend a larger fraction of their R&D on basic research are more productive, have a higher level of output relative to their other measured inputs, including R&D capital, and this effect has been relatively constant over time." Griliches finds, just as Mansfield did six years earlier in a smaller sample, that the private return on investment in basic research is several times higher than that on applied or developmental research.

The conclusion is probably warranted that private rates of return on R&D are higher than those on other private investments—if we are to believe a number of respectable studies. In Canada in particular this seems to be the case. We shall take up the subject of *social* returns to R&D when we discuss subsidies in a later chapter.

A mystery remains and casts doubt on the above conclusion. If ROI on R&D is so high, why has it not been driven down over time by profit-maximizing corporations who will try to equate returns to production factors on the margin? Such an occurrence has been found by Griliches with respect to the use of chemical fertilizers in U.S. agriculture.[9] And, of course, why do we not see money spent on R&D rise over time in Canada as a consequence of this profit-maximizing behaviour of private firms?

A tentative answer may be a high, but non-measured, level of risk connected with innovativeness. Profitability, after all, has two dimensions; the rate of return and its variability or risk. And yet, as we indicated in the previous chapter, both the New York and Toronto stock exchanges put a premium on the research intensity of member firms. Since stock market valuation does take account of risk, the tentative

---

8   Edwin Mansfield, "Basic Research and Productivity Increase in Manufacturing," *American Economic Review*, December 1980, pp. 863-873.

9   Zvi Griliches, "Research Expenditures, Education and the Aggregate Agricultural Production Function," *American Economic Review*, December 1964, pp. 963-974. Yet the same Griliches remarks, 22 years later, again in an *American Economic Review* article, that one's response to this (mystery) depends on one's views as to the prevalence of equilibria in the economy. See Griliches 1976, *op. cit.*

answer to the mystery of too high returns to industrial research remains so far unanswered.

Under certain circumstances, it is useful to enlarge the profitability perspective and examine other outcomes of innovation that are important to the firm. In a study of 43 European and 28 Japanese multinationals Kotabe found that a combination of market share, sales growth rate, and pretax profitability, called "performance" of a new product was strongly related to that product's innovation "magnitude."[10] In general, studies examining the success or otherwise of innovations, for reasons already mentioned, tend to concentrate on less "hard" measures than on profitability, as will be seen in the following discussion which turns to the question:

What makes innovations successful?

Having established that innovative endeavour can often lead to a profitable outcome, management is likely to ask next what particular paths to innovative products or processes could improve the chances of their success. This question can be put more simply as "what are the determinants of successful industrial innovations?" What comes immediately to mind is that such determinants can be classified as being either internal to the firm or largely influenced by the firm's environment. Management literature naturally concentrates mostly on the internal determinants, since these are the factors the firm can more easily manipulate. Economic investigations throw some indirect light on the role of environmental influences.

Understanding the importance of internal and environmental (or external) determinants of innovation is important. If there is a systematic pattern, a constellation of factors that add up to successful innovation, then this pattern could provide guidance both to management in its quest for increased profitability and to public authorities anxious to direct support to worthwhile projects with minimum risk. A quarter of a century of investigations has yielded some valuable results but their

---

10   Masaaki Kotabe, "Corporate Product Policy and Innovative Behaviour of European and Japanese Multinationals: An Empirical Investigation," *Journal of Marketing*, April 1990, pp. 19-33.

implications should be tempered by the realization that a panacea in this area is as difficult to come by as the stone of philosophers upon which, it is said, rests the success of the Japanese corporation.

Roughly speaking, there are basically two ways to examine the puzzle of innovation. The first zeros in on the innovative product or process, and considers the corporate management context as well. The second downplays the role of the individual innovation and concentrates on the firm that carries out innovative activities over some period of time. Both approaches try to single out the groups of factors that determine the commercial success of innovation. Perhaps the chief difference between the two is in their view of what constitutes commercial success. The innovation-centred approach tries to match the costs and benefits that are closely connected with the new product or process. The firm-centred or system-oriented approach considers innovative products or processes as just one of the constituent outcomes of a company's complex output, where the company's market success must be appraised in a long-run perspective.

The SAPPHO Project—Individual Innovations

The best known series of investigations concerning the managerial aspects of individual innovation was undertaken at Sussex University in the United Kingdom under the project name SAPPHO—Scientific Activity Predictor from Patterns with Heuristic Origins. The project was designed as a systematic attempt to discover differences between successful and unsuccessful product and process innovations in the chemical and scientific instrument industries of Western Europe and the United States over the post-war period to about 1970.[11] The technique employed was one of paired comparisons, in which the two innovations, one successful and the other not, competed for the same *market* while not being technically identical. The criterion for success was

---

11   Christopher Freeman, *Economics of Industrial Innovation*, London: Penguin, 1974, Chapter 5; Roy Rothwell et al., "SAPPHO Updated: Project SAPPHO Phase 2," *Research Policy*, November 1979, pp. 258-291.

commercial; a successful innovation was defined as one that obtained a worthwhile market share and/or profit.

Note that "worthwhile market share" and "worthwhile profit" are to some extent judgemental criteria; they have not been quantitatively defined in the study. The investigators themselves note that the overall success of an innovation must be measured by the total impact of the innovation on the innovating organization, and suggest that when share and profit figures give ambiguous results, "alignment with company strategy" could tip the scales of the success criterion. There remains the essential ambiguity inherent in accepting as success that which management defines as such; it is the non-removable bane of management studies that rely on interviews.

The comparisons were made against more than 120 criteria which at that time (1970) were believed to discriminate between success and failure in innovation. Some of these criteria are applicable to the innovation itself, some to the innovating firm. The flavour of the comparisons made in the 43 pairs of innovations (22 in chemicals, 21 in scientific instruments) is indicated by the partial results presented in table 10. Table 11 lists the individual innovation pairs.

One-third of the 120-odd criterion variables proved to be statistically significant discriminators, at the 95 percent confidence level, between success and failure. These variables were also assembled into ten aggregate indices. Five of these together accounted for virtually all of the observed differences between success and failure. They are listed in table 12. The index variable "marketing," which correctly classifies 83.7 percent of the innovations, is a composite measure of the marketing effort deployed by the innovating organization and consists of the following individual variables:

- the innovation was part of a general marketing policy
- attention given to publicity and advertising steps taken to educate users
- sales effort a major factor
- a marketing decision (the observation of a need) rather than a production decision (addition to a product line created by new technology)

Table 10: Five Individual Variables that Discriminate Most Strongly Between (S)uccess and (F)ailure of 43 Pairs of Chemical and Scientific Instruments Innovations

Ranking[a]	Variable	S>F	S=F	S<F
1	Successful *firms* understand user needs better (UN)	33[b]	10	–
2	Successful *innovations* have fewer aftersales problems (RD)	31	13	–
3	Successful firms employ larger sales efforts (MKTG)	22	21	–
4	Successful innovations have fewer bugs in production (RD)	25	17	1
5	Successful firms have better coupling in specialized areas (COMM)	23	19	1

[a] Probability of chance occurence in all five results is less than 1%.

[b] In 33 of the 43 innovation pairs user needs were better understood by successful firms; in the remaining 10 successful or unsuccessful innovations firms understood (or misunderstood) such needs equally.

Sources: Christopher Freeman, *Economics of Industrial Innovation*, London: Penguin, 1974, Chapter 5; Roy Rothwell et al., "SAPPHO Updated: Project SAPPHO Phase 2," *Research Policy*, November 1979, pp. 258-291.

In a similar vein, "R&D strength" is a measure of the performance of the development work on the innovation; "user needs" a measure of the efficiency with which market research or other procedures have established the precise requirements of the customer; "communication" a measure of the effectiveness of the innovating organization's communication network with the outside scientific and technical community. Finally, "management strength" is a measure of coordination. Four of the five index variables are represented by one of their individual

Table 11: List of 43 Innovation Pairs

Chemicals	Scientific Instruments
1. Accelerated Freeze-Drying of Food (Solid)	1. Amlec Eddy-Current Crack Detector
2. Acetic acid Preparation	2. Atomic Absorption Spectrophotometer
3. Acetylene from Natural Gas	3. Automatic clinical analysers I
4. Acrylonitrile I	4. Automatic clinical analysers II
5. Acrylonitrile II	5. Automatic clinical analysers III
6. Acrylonitrile III	6. Automatic clinical analysers IV
7. Acrylonitrile IV	7. Digital Voltmeters
8. Ammonia Synthesis	8. Electromagnetic Bloodflowmeter
9. Caprolactam I	9. Electronic Checkweighing I
10. Caprolactam II	10. Electronic Checkweighing II
11. Caprolactam III	11. Electronic Checkweighing III
12. Caprolactam IV	12. Electronic Checkweighing IV
13. Ductile Titanium	13. Foreign Bodies in Bottles Detector
14. Extraction of Aromatics	14. Milk Analysers
15. Extraction of n. Paraffins	15. Nuclear Magnetic Resonance Spectrometers
16. Glass Ceramics	16. Nucleonic Thickness Gauges
17. Hydrogenation of Benzene to Cyclohexane	17. Optical Character Recognition
18. Methanol synthesis	18. Particle Counters
19. Oxidation of Cyclohexane	19. Roundness measurement
20. Phenol synthesis	20. Scanning Electronic Microscope
21. Steam Naptha Reforming	21. X-Ray Microanalyser
22. Urea Manufacture	

The sample is international, and includes innovations from the United Kingdom, the United States, Germany, Italy, France, the Netherlands, Denmark and Switzerland.

Source: Same as table 10.

"sub"-variables in Table 10, where the mnemonic abbreviations in brackets designate user needs, R&D, marketing and communications, respectively.

As can be expected, there are strong correlations between the scores of the five index factors. On average, the successful firm outperforms the unsuccessful firm in all five areas of competence for the given innovation. As the SAPPHO authors point out, one must look to multi-factor explanations for success and failure in industrial innovation. Nevertheless, certain fairly simple conclusions do emerge from the study. Innovation is a coupling activity comparable to the blades of a pair of scissors. One blade represents the recognition of a potential market or in-house application for a new product or process. The other blade stands for unfolding technical knowledge, outside or inside the firm. The blades meet by matching the technical possibilities and the market needs.

The SAPPHO results unambiguously support the belief that *firms that had a successful innovation paid more attention to the market than unsuccessful companies.* The discriminating performance of the composite variables **marketing** and **user needs** attests to this. A strong in-house R&D (R&D **strength**) as well as good **communications** with outside technological developments in the general area relevant to the innovation proved to be another two of the five strongest discriminants between failure and success. As one of the SAPPHO project leaders remarks, the test of successful entrepreneurship and good management is the capacity to link together market and technology, by combining the two flows of information.

And, indeed, the fifth statistically significant discriminant composite index in Table 12 is designated **management strength.**

Screening for Successful Innovations

Building on SAPPHO and other studies Cooper took the next step and proposed a new-product model for predicting success or failure.[12] The

---

12   Robert G. Cooper, *A Guide to the Evaluation of New Industrial Products for Development*, Montreal: Industrial Innovation Centre, 1982.

Table 12: Five Index Variables that Discriminate Most Strongly between Success and Failure of 43 Innovations

Index Variable	% of pairs for which S>F		
	Total Sample (43)	Chemicals (22)	Instruments (21)
Marketing (MKTG)	(1) 83.7	(2) 72.7	(1) 95.2
R&D Strength (RD)	(2) 76.7	(1) 81.8	(4) 71.4
User needs (UN)	(3) 74.4	(2) 72.7	(2) 76.2
Communication (COMM)	(4) 69.8	(5) 63.7	(2) 76.2
Management Strength	(5) 65.1	(4) 68.2	(2) 61.9

Source: Source: Christopher Freeman, *Economics of Industrial Innovation*, *op. cit.*, pp. 258-291.

model was based on a survey sample of 186 industrial product innovations of Canadian manufacturing firms. Answers to the question about background characteristics of successful and failed innovations were factor-analyzed and aggregated into 13 larger dimensions. The degree of success/failure was then regressed onto these dimensions of determinants, of which 7 proved statistically significant. The model could predict success or failure in 85 percent of the cases.

An example is given to illustrate the procedure. Previously published research suggested that the innovation project's newness to the firm would detract from expected success. The survey questions (variables) touched therefore on whether the expected customers were new to the firm, whether the competitors to be encountered were new, whether the required production process was new, etc. Replies all loaded onto the factor "newness to the firm." The score of this factor was one of the regressors. It proved significantly negative.

The next step was to suggest a screening model called NEWPROD which is used for an overall rating on a proposed new industrial product project. Based on subjective opinions of within-firm or consulting evaluations or, in other words, on expert information at a stage at which no heavy project expenses are incurred yet, the screening model is proving highly successful and is periodically verified against new results.

Appendix 4A reproduces the first pages of the model's rating form to convey the flavour of this successful Canadian product.

We dwelt on Cooper's NEWPROD model for three reasons. First, reasonably successful attempts are being made to translate inquiries about the determinants of innovation into operational means to help with innovativeness. Second, the model shows that R&D is only one of the many elements entering management's forecast of a successful new product or process. Third, in the 48 tested questions submitted to project evaluators, there is not a single one that refers to the government's potential support, either in the form of tax relief or in the guise of subsidy.

This last point leads us to consider what *external* influences may influence the success of innovations. Clearly, market demand is the most powerful, but we chose to categorize it under "internal" since it is so close to the firm's concerns. Technological opportunity or changing technological environment is a powerful stimulant of industrial innovation and has documented influence on R&D expenditures, but does not necessarily contribute to innovative success.[13] Similarly, the competitive structure of the industry has been subjected to innumerable investigations with contradictory results, but again with R&D expenditure rather than innovative success as the variable to be accounted for.[14]

Finally, there is government support. Studies under this heading tend to be of a certain age, but they also tend to unanimity: government policies, whether by direct subsidy, purchasing schemes, merger encouragement or tax alleviations to encourage innovation have been, at

---

13   The Zif *et al.* article (*op. cit.*) offers evidence in this respect.

14   Cooper's count of "many vs. few competitors" is distant from the industrial organization inquiries; he finds that the more competitive the field is, the less chance of a success.

best, ineffective.[15] Having examined at length aspects of individual innovations, we now turn to consider innovative firms.

The Louvain Studies—Innovation in Firms

In 1965 the Belgian Ministry of Science Policy asked a group of researchers at the Université Catholique de Louvain (UCL), led by Professor Ph. de Woot, to establish guidelines which would facilitate the allocation of public funds among enterprises asking for R&D support for specific projects.[16] The group undertook an extensive study of 96 firms representing 21 percent of Belgian firms engaged in industrial research and 69 percent of the total industrial R&D outlays. The principal hypotheses to be tested were that a firm's profitability increases with research intensity as well as with the size of the company, and that it is also influenced by the type of industry in which it participates. Profitability, however, as measured by a seven-year average rate of return on equity of the firm, did not prove to be statistically related to any of the hypothesized determinants.

What the researchers did find in the course of their survey interviews and data examination stretching over three years was, nevertheless, a definite, consistent pattern which it is instructive to compare with the SAPPHO findings:

1. While there is no discernible relationship between R&D outlays, or the number of innovations commercialized, and return on equity, R&D activity appears to be associated with higher profitability when

    a) innovation project selection is a joint responsibility of top management, marketing, production, and R&D representatives;

---

15   One reference here is sufficient, more will be forthcoming further on. Rubinstein et al., "Management Perceptions of Government Incentives to Technological Innovation in England, France, West Germany and Japan," *Research Policy*, 1977, pp. 324-357.

16   Ph. de Woot and H. Heyvaert, "Management stratégique et performance économique," *Economies et sociétés*, 13, 1979, pp. 509-37.

b) a systematic market analysis is conducted in step with technical development;

c) explicit criteria are used to re-assess R&D projects with a view to abandonment.

2. Profitable firms adhere to a systematic product policy, practice a balanced allocation of human resources among production, marketing, and R&D functions, have a better knowledge of their environment, and have their executives spend more time in long-range planning.

These results formed the basis of another wave of research that formally tested the basic hypothesis that the economic success of a firm is largely determined by the quality of its management strategy, which in turn is reflected in four principal areas: a definite product policy; an equilibrium between the principal functional areas of marketing, production, research, top management; an internal and external information system; and a systematic innovation policy. A sample of 12 Belgian research-intensive firms of various sizes was chosen from fast-and-slow technologically changing industries for a thorough investigation. The basic hypothesis was handsomely confirmed.

Adoption of Innovations

Given that most of the firm's technology is "received" rather than developed in-house, economists and management scholars have devoted a lot of attention to this second aspect of innovativeness. Its importance is attested to by the already cited work of Terleckyj in the U.S. and Postner and Wesa in Canada (Box 2). A vivid illustration of the importance of adoption (and so, *mutatis mutandis*, of diffusion) of technological innovation in Canada is given by Lucas.[17]

Lucas looked at data over the 21 years from 1963 to 1983 emanating from two groups of Canadian industries:

---

17   R.G. Lucas, "High vs. Low Technology: Assessing Innovation Efforts in Canadian Industry," *Canadian Journal of Administrative Sciences*, June 1986, pp. 121-145.

A	B
Steel making	Aircraft and aerospace
Pulp and paper	Communications equipment
Mining	Electrical equipment
Industrial chemicals	Plastics and resins

Group A can be designated as low in R&D intensity, group B as high. Suppose now that most of the technological innovation in B is aimed at new or improved products, in A at more efficient processes. This seems likely if we consider the standardized outputs of group A. Both groups of industries depend on continued technological competitiveness, but group A buys most of it in the form of new process equipment from suppliers: it *adopts* the process innovations offered to it. (Undoubtedly, much collaboration takes place between suppliers and customers in the course of adoption and adaptation).

The statistical measure of process innovation Lucas used is machinery and equipment net fixed capital formation, MENFCF, which represents the acquisition of new machinery and equipment over and above that required for replacement purposes. It is an *input* approximation to innovation adoption, just as R&D is an input measure of innovation creation.

Table 13 indicates that while the median *R&D* intensity of group B industries is (R&D/Sales) = 5.8 percent, the same ratio for group A is only one-tenth of this. On the other hand the "low-tech" group's median *process* intensity, as measured by the ratio (MENFCF/sales) is 3 percent, or three times that of the "hi-tech" group's. The table tells a story not only of the other face of innovativeness, namely adoption, but also of the fragile distinctions between hi- and lo-tech designations.

Success in Adoptions

Adoption of an innovation by a buyer, whether of industrial or of the consumer variety, is the necessary consequence of its sale by the creator-innovator. In that sense, what makes adoptions happen is what makes innovations commercially successful. As discussed in the preceding

Table 13: Median R&D and Equipment Acquisition Sales Intensities Per Year, 1963 to 1983

	Group A industries	Group B industries
	Steel, Pulp & Paper Mining, Ind. Chemicals	Aerospace, Communications, Electrical, Plastics
Median per year sales (million $)	4,411	1,243
M.P.Y. R&D (million $)	23	72
M.P.Y. MENFCF (million $)	117	12
R&D/sales	0.005	0.058
MENFCF/sales	0.027	0.010

Source: G. Lucas, "High vs. Low Technology: Assessing Innovation Efforts in Canadian Industry," *Canadian Journal of Administrative Sciences*, June 1986, pp. 121-145.

———

chapter and in the preceding sections, the factors which make for speedier adoptions are manifold, but few appear to be government policy-influenced, unless, of course, one counts the framework policy stance of encouraging competition as a direct innovation policy.

Two aspects of innovation adoption, the other face of the coin of innovativeness, are worth mentioning as "determinants of success." Increasingly, management analysts observe and recommend greater cooperation between suppliers and customers (subcontractors and con-tractors, manufacturers and distributors, etc.). This applies a *fortiori* to new technology. In many instances it is nowadays difficult to make a distinction between innovators and adopters, both sides being respon-sible for the development of new products and processes. Box 7 repro-duces an exhibit from a recent Canadian National Centre for Management Research and Development publication. Therein Profes-

Box 7: Generic Decision Stages in the Development and Adoption Processes

Developing (Selling) Organization	Stage	Adopting (Buying) Organization
Analysis and definition of the internal organizational need and strategic rationale for developing new technology/product/market scenarios	Problem Recognition	Analysis and definition of the internal organizational need and strategic rationale for adopting new technologies/products.
Analysis and definition of needs, payoffs, and risks of market segments in adopting a new product	Need Analysis	Analysis and definition of needs, payoffs, and risks of adopting a new product.
Analysis and definition of the choices between alternative product performance dimensions and physical product features for different market segments	Product Concept	Analysis and definition of ideal performance dimensions and physical features for the needed product.
Analysis, definition, choice, and linkage of alternative physical technologies for the product concept	Technology Choice	Analysis, definition, choice and linkage of alternative physical technologies-in-use for the product concept.
Analysis and definition of the financial viability of developing alternative new technology/product/market scenarios	Financial Analysis	Analysis and definition of the financial viability of adopting alternative technologies/products.
Analysis, definition, and detailed translation of technologies and product concept into physical products for target market segments	Product Design	Analysis and detailed definition of the required physical and performance attributes for a new product.
Analysis and planning of production process, sourcing, and logistics to produce product units	Production Sourcing	Analysis and planning of sourcing, search for alternative technologies, products and suppliers.

Box 7 (continued)		
Developing (Selling) Organization	**Stage**	**Adopting (Buying) Organization**
Commitment of financial and human resources to produce and market product units	Unit Commitment	Commitment of financial resources to purchase product units and choice of suppliers.
Testing of customer process implementation and integration of use of the new technology/product and modification of the product	Use Implementation	Process implementation and integration of use of the new technology/product, analysis and evaluation of effectiveness and satisfaction, analysis of changes necessary, and repurchase.

Source: Roger A. More, *Improving the Adoption of Technology: A New Framework*, London, Ontario: National Centre for Management Research and Development, 1990, Exhibit 2, p. 7.

sor More proposes an integrated model of developer-adopter cooperation for successful innovation.

Let us mention briefly, as the other aspect, the success of governments in speeding up adoption, either of new technology emerging from its own labs or of private sector innovations. The latter endeavour finds expression in the funding and general support of innovation centres across Canada; the former in the transfer of technology from governments to the private sector, which has been one of the first mandates of the National Research Council. In part it is addressed by IRAP (Industrial Research Assistance Program), started in 1962. By fiscal year 1985/86 annual expenditures thereon reached about $40 million. From then on IRAP absorbed PILP, the Program for Industry/Laboratory Projects which started ten years earlier. PILP was even more focused on transfer, particularly to the small and medium enterprise. Together the two programs were spending over $50 million in fiscal 1987/88.

In his penetrating review of the effectiveness of certain federal government programs in support of technological innovation, Tarasofsky of the Economic Council of Canada stated that with respect to IRAP: "the information formally required from applicants is utterly incapable of permitting a rational judgment as to whether the project warrants subsidization."[18] He held a similar opinion of PILP.

A subsequent report by a governmental advisory committee found that "...a major shortcoming relates to the intramural capacity (of the federal government) for technology transfer. Although successive governments have made the commercialization of technology a high priority of science policy, the actual record is poor."[19] It also cites with approval the United States Federal Technology Transfer Act of 1986 which provides incentives to federal government laboratories to transfer technology and whose consequences have been favourably reviewed in a 1989 Secretary of Commerce report.

Implications for Innovation Support Policies

The first purpose of this chapter was to emphasize that R&D is but one of the activities that management must undertake to innovate and to keep the firm competitive. As we have seen from the description of the SAPPHO and Louvain studies, the task of coordinating the innovative effort, of bridging the gap between the firm's technological supply and the market's demands is complex and requires high managerial skills.

Neither of these two studies or the normative NEWPROD model alluded to government support. However if there is to be direct support to private firms in the form of grants or studies, the policy implications here are that it should not be confined to the easily-defined target of R&D activity, nor should it go to firms with weak management. But the identification of good management in firms with little track record is

---

18 Abraham Tarasofsky, *The Subsidization of Innovation Projects of the Government of Canada*, Ottawa: The Economic Council of Canada, 1984, p. 58.

19 National Advisory Board on Science and Technology, *Revitalizing Science and Technology in the Government of Canada*, Ottawa, November 2, 1990. Released April 1991, p. 103.

difficult. Attractive projects, or even strategic plans for the innovation in question, require continuous monitoring, an activity difficult to achieve in public bureaucracies.

Beyond administrative difficulties there looms an issue which this writer, at least, has not been able to fully resolve: given that successful innovations are ultimately the result of strong management, public funds to stimulate innovation should go to well-managed firms. But well-managed enterprises are by definition innovative when market demands innovation, and do not need public funds to make them so. Should, therefore, public funds be expended on support to industrial innovation?

But let us not conclude on a "passive" note, and let us take up the point about subsidies to non-R&D innovative activities.

As we already mentioned, the risk of failure of an innovative product in its commercial stage is at least as high as the risk that the research and development effort devoted to the product will not reach successful completion. Similarly, the investment in the commercial side of the innovation, namely in manufacturing start-up and marketing start-up, exceeds by a large margin the investment in R&D.[20] Thus the investment required to launch it and the accompanying risk are much greater in the typical innovative venture than the research outlays. This can be represented by a specific version of the product life cycle model, tailored to reflect the dangers facing innovative firms in industries with rapidly changing technologies.

In Figure 11 the development time, D_p, of the innovation is relatively long and costly, the introduction/growth time, I/G, is long, while the maturity period M is short as the market changes rapidly under the impulse of competitive innovations. (D stands for decline.) On a net present value basis, using the firm's capital cost as a discount rate, the investment project is probably barely, if at all, profitable.

The point is that Canadian government granting agencies prefer to subsidize the R&D part of innovative projects, and that Canadian tax

---

20 The De Melto et al. Economic Council of Canada study (footnote 4, chapter 3) covers only industrial products where marketing costs tend to be low. Even so, R&D constitutes only 42 percent of the total outlays.

laws strictly exclude from its favourable treatment of research and development anything that goes beyond pilot plant outlays, and that more and more federal support to innovation is channelled through tax concessions rather than direct grants. The cumulative effect of these policies is that innovative enterprises are not supported in their most perilous periods. If subsidization of innovation is on the cards, then the whole innovation investment should be eligible for favourable treatment—not just R&D.

Figure 11: An Unprofitable Product Life Cycle

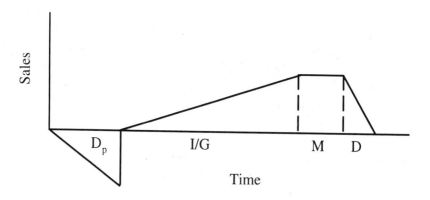

Source: Arieh Goldman and Eitan Muller, "Measuring Shape Patterns of Product Life Cycles: Implications for Marketing Strategy," an unpublished paper, August 1992, reproduced in Philip Kotler's *Marketing Management*, Englewood Cliffs, N.J.: Prentice Hall, 1984, p. 359.

Appendix 4A

NewProd Screening Model:
Rating Form

Evaluator[a] _____ of
_____ evaluators
Project I.D.[a] _____
Project Name[a] _____

INSTRUCTIONS

Please read instructions carefully before beginning.

The NewProd Screening Model is a decision model that combines subjective opinion from a number of evaluators to yield an overall rating on a proposed new product project.

As one of these evaluators, you are asked to provide your thoughts or ratings on a number of characteristics of the project identified above (see Project Name, above).

Please read each statement listed below. Do these characteristics describe the project? Indicate your degree of agreement or disagreement by circling a number from zero (0) to ten (10) on the scale to the immediate right of each statement:

Here:

> 0 means strongly disagree
> 10 means strongly agree
> and numbers between 0 and 10 indicate various degrees of agreement or disagreement.

Please provide a rating for every statement even though you may not be certain about your answer.

You are also required to indicate how certain or confident you are about each of your responses. Do this by writing a number from 0 to 10 in the column to the far right on each page, headed CONFIDENCE.

Here:

> 0 means very low confidence in answer; highly uncertain
> 10 means total confidence; highly certain
> and numbers between 0 and 10 indicate varying degrees of confidence.

---

[a] to be filled out by model administrators.

Source: Robert G. Cooper, *A Guide to the Evaluation of New Industrial Products for Development*, Montreal: Industrial Innovation Centre, 1982.

Appendix 4A (Continued)

Resources Required	Strongly Disagree	Strongly Agree	Confidence (0 to 10)
1. Our company's financial resources are more than adequate for this project.	0 1 2 3 4 5 6 7 8 9 10		_____
2. Our company's R&D skills and people are more than adequate for this project.	0 1 2 3 4 5 6 7 8 9 10		_____
3. Our company's engineering skills and people are more than adequate for this project.	0 1 2 3 4 5 6 7 8 9 10		_____
4. Our company's marketing research skills and people are more than adequate for this project.	0 1 2 3 4 5 6 7 8 9 10		_____
5. Our company's management skills are more than adequate for this project.	0 1 2 3 4 5 6 7 8 9 10		_____
6. Our company's production resources or skills are more than adequate for this project.	0 1 2 3 4 5 6 7 8 9 10		_____
7. Our company's sales force and/or distribution resources and skills are more than adequate for this project.	0 1 2 3 4 5 6 7 8 9 10		_____
8. Our company's advertising and promotion skills and resources are more than adequate for this project.	0 1 2 3 4 5 6 7 8 9 10		_____

Appendix 4A (Continued)

An Example of a New Product Project

Solidstate A.C. Varispeed Motor #127			
Question	Mean weighted score	Standard deviation	Mean confidences
1. Financial resources	7.4545	2.0165	7.3333
2. R&D skills	7.9412	2.2088	5.6667
3. Engineering skills	7.6667	2.2111	5.0000
4. Mkting research skills	7.4286	1.3997	4.6667
5. Management skills	7.3000	0.9539	6.6667
6. Production resources	7.0400	0.8237	8.3333
7. Salesforce resources	7.3333	0.9428	7.0000
8. Advertising skills	5.6250	2.4969	5.3333
9. Product innovation	8.7143	0.9583	9.3333
10. Hightech product	7.7368	2.0734	6.3333
11. High per unit price	7.2000	3.1241	6.6667
12. Mechanical complexity	8.8889	0.9938	6.0000
13. Idea from mkt-place	7.0000	1.1421	7.6667
14. Clear product specs	5.3636	0.4811	7.3333
15. Clear technical aspects	3.0000	1.0954	8.3333
16. Custom product	0.0	0.0	10.0000
17. Defend mkt share	0.0	0.0	10.0000
18. Relatively expensive	6.7273	1.2498	7.3333
19. New potential customers	1.0870	2.0624	7.6667
20. New product class	3.1176	3.7711	5.6667
21. Never made/sold this type	4.0000	4.7434	4.0000
22. New production process	1.2632	2.9171	6.3333
23. New R&D	1.9000	2.4062	6.6667
24. New distribution systems	3.3333	3.9441	6.0000

Solidstate A.C. Varispeed Motor #127			
Question	Mean weighted score	Standard deviation	Mean confi- dences
25. New advertising	2.6154	4.0103	4.3333
26. New competitors	2.6923	3.4055	4.3333
27. Competitive uniqueness	10.0000	0.0	10.0000
28. Competitive superiority	10.0000	0.0	9.3333
29. Reduced customer costs	8.8235	3.2219	5.6667
30. Improved cstmr prdctivity	10.0000	0.0	8.0000
31. Higher quality	10.0000	0.0	7.6667
32. Higher price	8.1538	2.8513	4.3333
33. Market innovation	9.2857	0.4518	9.3333
34. Mass market	8.6923	1.5385	8.6667
35. High customer need	8.9583	0.7895	8.0000
36. Only potential demand	0.0	0.0	5.3333
37. Large market $ size	7.6316	1.6610	6.3333
38. Fast-growing market	8.5000	1.5000	4.6667
39. High product homogeneity	6.2500	4.4371	4.0000
40. High market competition	1.3529	2.4720	5.6667
41. Intense price competition	1.3500	1.5898	6.6667
42. Many competitors	0.3182	0.4658	7.3333
43. One dominant competitor	0.2800	0.4490	8.3333
44. High cstmr loyalty to cmp	2.2632	2.1236	6.3333
45. High cstmr satisf w/comp	2.8400	1.8040	8.3333
46. Frequent newprod intros	2.0000	2.1602	5.0000
47. Dynamic market	2.3529	2.2476	5.6667
48. Many govt regulations	2.0833	2.4650	4.0000

Chapter 5

Le Mal Canadien?

"L'impression se répand qu'une malédiction particulière pèse sur la France... L'idée se crée, comme à d'autres périodes difficiles de notre histoire, qu'il existerait un mal français , une sorte de maladie chronique, insaisissable et spécifique. . . . Il n'y a pas de mal français."

Jean-Jacques Servan-Schreiber[1]

Introduction

A S THIS CHAPTER WAS BEING WRITTEN, the annual World Competitiveness Report put Canada in fifth place among the world's 10 industrialized countries. Among the 330 criteria that enter the judgment some apparently touch on science and technology. "Canada ranks second in infrastructure, fifth in people and sixth in the extent to which government policies affect competitiveness. *It shows poorly in science and technology.*"[2]

---

1   From *L'express* editorial, August 23, 1976, cited in Alain Peyrefitte, *Le mal français*, Paris: Plon, 1976.

2   *Globe and Mail*, June 20, 1991, B8. Since then Canada has been "demoted" to eleventh place. One of its worst scores is, however, still the one on

What solid evidence do we have that Canada is doing poorly in technology—and so, presumably, in industrial innovation? Or, at least, what weak evidence and interpretation do we have? Now it may well be that on some absolute scale, present in the mind of sophisticated technocrats and gullible media, Canada is not doing well enough. Our measure of that, however, should be both time-series or cross-sectional numerical evidence. In less technical jargon, we should look at developments over time in innovation proxies for Canada, and at recent international comparisons between Canada and other countries. But because of what is obfuscatingly called "aggregation problems," we should also be hesitant about whole-economy, whole-country measures and prefer, where available, individual industry or sectoral appraisals.

Finally, it should not be forgotten that even if weaknesses in innovation or competitiveness are discovered, industrial policy measures to remedy such could prove either irrelevant, unpolitic, or inefficient. With this reminder, let us turn to the evidence on and opinions about *le mal innovatif canadien*. The narrative can be structured along the lines of Figure 1 in which input and output proxies of innovativeness are listed.

The R&D Input Proxy
Economy-Wide Research Intensity

As stated, we look at the evidence both over time for Canada, and between Canada and other economies. The statistic most often employed here is the GERD/GDP, or the ratio of gross expenditures on R&D to gross domestic product. Table 14 gives the relevant figures. Over the eight years 1983 to 1990, total R&D expenditures in constant 1981 dollars grew at a respectable 4.1 percent per year. But that did not improve the 1.3 to 1.4 GERD/GDP ratio which is forever the target of complaints and testimony to failed government objectives.[3]

---

technology, *Globe and Mail*, June 22, 1992, p. B2.

3   Over the last 15 years federal governments, both Liberal and Conservative, tried in vain to jawbone the private sector into helping to lift the ratio over 1.5 and up to 2.5.

Table 14: Gross Domestic Expenditures on R&D (GERD) in Current and 1981 Dollars and in Percentages of the Gross Domestic Product, 1983-1990

	1983	1984	1985	1986	1987	1988	1989	1990
GERD in current	5,348	6,015	6,709	7,220	7,542	8,058	8,578	9,097[a]
and in 1981 dollars (millions)	4,687	5,110	5,558	5,841	5,847	6,000	6,085	6,231
GERD/GDP	1.32	1.35	1.40	1.43	1.37	1.33	1.32	1.34

[a]Estimate

Source: Industry, Science and Technology Canada, *Selected S&T Statistics 1990*, Ottawa: February 1991.

This rough and ready ratio is most often used for international comparisons, as in Figure 3, wherein Canada is shown to be attaining less than half the ratios for Sweden, Germany, the U.S., and Japan. The patent absurdity of this comparison appears as soon as the composition of GERD is examined. Consider figure 12. There the black rectangles show the proportion of the GERD/GDP ratio that goes to defence R&D. For Canada, it is minuscule. In fiscal year 1986/87 it amounted to $221 million, or 8.6 percent of the federal budget devoted to R&D. These public R&D expenditures on defence approximated 3 percent of GERD in 1987. Compare this to about one-third of GERD devoted to defence in the U.S. and the substantial parts of GERD going to the same objective in France, the U.K. and Sweden.

Less visually easy, but more statistically convincing in this respect is Table 15 which lists 14 countries in order of increasing defence burden. Rank correlations are calculated between defence and overall research intensities, and confirm their positive links.

The picture is clear: countries that take upon themselves heavy defence responsibilities will necessarily incur heavy research expenditures. Such expenditures are a function of political commitments, not of innovative inclinations.

Sectoral Research Intensities—Macrosectors

The composition or the structure of an economy, as well as its defence responsibilities, matter in influencing research intensity. As of now the technological opportunity for investment in research is still the highest in the manufacturing sector, though service sectors such as software design are raising their research intensity. In the late 80s in Canada manufacturing R&D accounted for about 79 percent of BERD, the Business Enterprise Sector Research and Development expenditures.

However, the manufacturing sector in Canada is responsible for only 20 percent of GDP, while the 1987 or 1988 figures for Germany, Japan, and Sweden are on the order of 32, 29 and 31 percent respectively. It is thus natural that just on this account, Canada's economy-wide

Figure 12: Composition of GERD/GDP, Selected Countries, 1987

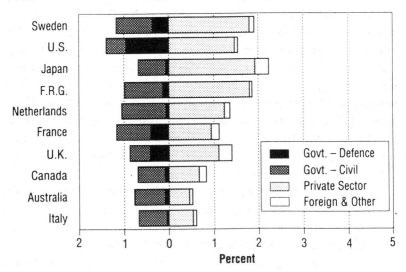

Source: *ISTC, Science and Technology Economic Analysis Review*, Ottawa: March 1990, Chart 2-3.

intensity is not fairly comparable to that of economies in which the manufacturing sector plays a larger role. A truer picture emerges in Table 16: Canada's weak showing in the GERD/GDP aggregate leagues improves somewhat after proper adjustment. A difference still remains, nevertheless.

Table 15: A Comparison of Defence and Economy-wide Research Intensities in Some OECD Countries, 1986

Country	Defence/GNP (in %)	GERD/GNP (in %)	Rank GERD/GNP
Japan	1.0	2.79	3
Austria	1.3	1.32	12
Finland	1.7	1.67	9
Denmark	2.1	1.32	12
Canada	2.2	1.44	11
Italy	2.3	1.14	14
Sweden	2.9	2.94[a]	1
Belgium	3.1	1.64	10
Germany	3.1	2.78[a]	4
Netherlands	3.1	2.22	7
Norway	3.2	1.70[a]	8
France	3.9	2.23	6
U.K.	5.0	2.34	5
U.S.A.	6.7	2.92	2

$r_s = 0.523$ (significant at 5%)
without Japan, $r_s = 0.817$ (significant at 1%)
[a]Value interpolated from adjacent years

Sources: United Nations, *Human Development Report,* New York: Oxford University Press, 1990, 163; OECD, *Main Science and Technology Indicators,* 1990-2, Paris, 1991, Table 5.

Table 16: International Comparison of Business Enterprise R&D Expenditures (BERD) as a Proportion of Gross Domestic Product: 1987

Country	BERD as a percent of total GDP (%)	Ratio of manufacturing industries' GDP to total GDP (%)	BERD as a proportion of domestic product of industry (DIP) (%)
Canada	0.77	19.8	1.08 (1986)
Australia	0.45	17.5	0.47
Denmark	0.79	18.4	1.24
Finland	1.02	22.1	1.44
Norway	1.13	15.1	1.55
Sweden	2.00	30.5	3.01
France	1.35	20.8	1.80
Germany	2.06	31.5	2.59
Japan	1.89	29.2 (1988)	2.12
U.K.	1.52	22.6	2.15
U.S.A.	2.11	22.4	2.40

Sources: OECD, *Main Science and Technology Indicators*, 1990-2, Paris, 1991, Tables 24 and 25, and *OECD Economic Surveys*, Various Countries, 1988-1990, Paris, 1990.

To see if this difference has narrowed over time, consider table 17 which lists the research intensity changes in the business enterprise sector for the G-7 countries. On this comparison Canada's BERD/GDP ratio has grown respectably both over time and *vis-à-vis* its big-time partners.

Let us finally speculate, still within the section on macrosector comparisons, on what Canada's big-sector research spending would be like if Canadian research intensity were governed by the average OECD sectoral composition of GDP. The formula for the simulated expenditure is

$$Simulated\ \$R\&D_{Sector,CDN} = \frac{\$R\&D_{Sector,\ CDN}}{\$GDP_{Sector,\ CDN}} \times \frac{\$GDP_{Sector,\ OECD}}{\$GDP_{Total,\ OECD}} \times \$GDP_{Total,\ CDN}$$

The sectors that would seem "naturals" are agriculture, mining, manufacturing, and services. However, statistics on the first and the last sectors are not abundant for the major OECD countries. In a previous volume, figures from 1975 indicated that Canada spent well above the OECD average on research in agriculture and mining and about "par" in the service sector.[4] The substantial shortfall was in manufacturing R&D.

Table 17: BERD/GDP Intensities and Growth, 1979 and 1990

Country	BERD/GDP 1979	BERD/GDP 1990	Annual Growth in %
Canada	0.46	0.70	4.74
France	1.06	1.40	2.92
Germany	1.66	2.00	1.86
Italy	0.43	0.70	5.71
Japan	1.20	1.98	7.18
UK	1.39	1.40 (1988)	0.08
USA	1.55	1.90	2.05

Source: Luc Soete, Background paper for the OECD conference on global technology, Montreal: January 1991, p. 36.

---

4   Kristian S. Palda, *Industrial Innovation: Its Place in the Public Policy Agenda,* The Fraser Institute, 1984, p. 86. The OECD countries: Austria, Belgium, Denmark, Finland, France, Germany, Italy, Japan, Netherlands, Norway, Sweden, UK, US.

Research Intensities in Individual Industries

Industry-specific reasons may account for a very large part of the difference between Canada's economy-wide research intensity and that of its industrially-advanced OECD partners. While we have already mentioned defence responsibilities and the overall economic structure of a country as important factors, on the individual industry level two important influences are prominent: the foreign-firm or multinational presence, and government intervention, either in the form of subsidy or of regulation. Only an industry-by-industry approach can draw out the particular circumstances determining industrial research intensity. This writer has been one of the first to advocate individual-industry comparative analysis across OECD economies and in 1992 the Science Council of Canada with Industry, Science and Technology Canada is publishing individual industry comparisons of a detailed nature.

For example, in the automotive sector, which is the largest apparent R&D under-spender, an explanation for the "shortfall" may be the integrated nature (read U.S. multinational presence) of the industry between Canada and the U.S. which, for Canada, results in low R&D-intensive assembly and higher R&D-intensive component manufacture.[5]

The analytical way to go about international industry comparisons of research intensities in individual three- or four-digit level industries was developed by this writer and a colleague in a background study for the 1983 Economic Council's *The Bottom Line*.[6] The principal idea is that a particular, reasonably narrowly defined industry faces a similar set of environmental influences and responds to them in its pursuit of profit in a similar manner, no matter in which of the advanced industrial countries it is located. Thus, for instance, the decision to allocate investment funds to research and development by the Canadian pulp and

---

5   ISTC, *Science and Technology Economic Analysis Review*, Ottawa: March 1990, p. 6.

6   K.S. Palda and B. Pazderka, *Approaches to an International Comparison of Canada's R&D Expenditures*, Economic Council of Canada, Ottawa: 1982 and Economic Council of Canada, *The Bottom Line*, Ottawa: 1983.

paper industry will be governed largely by considerations similar to those relevant to Swedish paper firms.

If a model based on this reflection can account for a good part of variation in R&D outlays or R&D intensity (R&D/Sales) in one industry across a number of countries over a number of years, it will have captured the essence of the response of that industry's R&D spending decisions to the stimuli and impediments in its past and present environment. In order to evaluate the adequacy or sufficiency of the *Canadian* industry's research intensity, the following steps are undertaken:

1. For a given industry, such as pulp and paper, assemble data on R&D spending and its possible determinants for as many OECD countries and for as many years as possible; exclude Canada.

2. Estimate by multiple regression the relationship

[$RD/$Sales] = f[Determinant variables $_{i,t}$]
  where country = i; year = t

   The result is the *average* relationship between research intensity and its causal influences for a given industry in a number of OECD countries.

3. Insert the Canadian industry's values of the determining variables into the above equation and calculate the predicted value of Canadian research intensity for a given year.

4. Compare the predicted intensity with the realized research intensity of the Canadian industry in the given year. If the realized intensity falls short of the predicted one, there are grounds for supposing that this Canadian industry underinvests in R&D (and so in innovation?) *when conformity to the OECD norm is assumed to be the criterion of optimality.*

Table 18 shows the forecast and actual research intensities for two of the seven industries analyzed—paper and chemicals. As the table shows, the research intensity for paper was, on average, higher than the OECD's standard, while the chemical industry's intensity was lower. In this case the other OECD countries were France, Germany, Italy, Japan, Sweden, the United Kingdom and the United States.

The main advantage of this approach to the assessment of a Canadian industry's R&D spending adequacy is that it takes into account the variation of R&D determinants across the countries to be compared and, in a sense, holds it constant. Thus, for instance, an important influence on R&D spending, the degree of foreign ownership, is incorporated into the model and "neutralizes" the influence of invisible R&D transfers from multinational headquarters to subsidiaries. The main value of this model is diagnostic: if OECD norms are the relevant ones (and they are continuously being invoked in public debate), then this approach identifies industries that *may be* in need of R&D stimulus (or should *not* receive more of it). Furthermore, it points to those determinants which may be susceptible to government action.

This industry-by-industry international comparison can be envisaged as substantial progress toward shaping an innovation policy, *provided* that R&D intensity is accepted as the determining factor of

Table 18: Comparison of Forecast and Realized R&D Intensities($RD/$Sales) In the Canadian Paper and Chemical Industries, (Percent)

	1967	1969	1971	1973	1975	1977	mean	Forecast as % of realized (mean)
Paper								
Forecast	0.46	0.57	0.43	0.32	0.37	0.37	0.42	87
Realized Value	0.77	0.59	0.45	0.35	0.36	0.39	0.49	
Chemicals								
Forecast	2.40	1.95	1.99	1.79	1.95	1.84	1.99	126
Realized Value	1.74	1.55	1.42	1.30	1.15	0.98	1.58	

Source: K.S. Palda and B. Pazderka, *Approaches to an International Comparison of Canada's R&D Expenditures*, Economic Council of Canada, Ottawa: 1982.

innovativeness. It represents the first two steps in the following "global" approach to policy support for industrial innovation:[7]

1. Define and measure statistically the gap between optimal and realized levels of innovative performance in Canada.

2. Estimate statistically the relationships between the gap and its several identifiable determinants.

3. Classify the determinants of the innovation gap as either unreachable or manipulable by policy-makers.

4. Consider pros and cons of the manipulable options.

5. Give policy recommendations, including the possibility of no policy.

Of the seven industries analyzed, three had higher R&D intensities than the regression had foretold, four had lower intensities than could have been expected given OECD norms. The estimates or forecasts just described represent both an over-time and across-countries perspective. Recently Hanel used an "as-if" way, similar in spirit to the formula given on page 121, to estimate a gap between the R&D investment performance of a group of seven Canadian manufacturing industries and that of their counterparts in 8 OECD countries.[8]

Between 1979 and 1985 the growth in research expenditures of the Canadian group—electrical, chemical, transport equipment, metal fabricating, machinery, chemical-allied, others—was on the order of 172 percent over those 6 years. In the OECD group it was 131 percent. Nevertheless, the average total "shortfall" of the Canadian group remained at a substantial 51 percent.

What generally accounts for such a shortfall in individual (mostly) manufacturing Canadian industries? Let us look at the major reason, the presence of foreign subsidiaries in Canada.

---

7   Though not innovativeness, because considerations of adoption/diffusion are not present.

8   Petr Hanel, "Ecart technologique de l'industrie canadienne," in Michel Leclerc (ed.), *Les enjeux économiques et politiques de l'innovation*, Quebec: Presses universitaires du Québec, 1990.

Visible and invisible technology transfers

It is well documented that foreign-controlled firms in nearly all manufacturing sectors do a lot less R&D relative to sales than do domestically-controlled firms.[9] The seeming paradox of foreign subsidiaries being heavily represented in R&D-intensive industries—intensive, at least, on the "own-performed" definition—and yet being less R&D intensive than their Canadian counterparts is resolved when it is recalled that such subsidiaries have less need to undertake research locally since they can rely on access to R&D results generated by their affiliates abroad. Statistically this explanation can be confirmed if either or both of the following conditions hold:

1. R&D- or technology-related payments abroad of foreign subsidiaries are relatively higher than those of Canadian-owned firms and most of these payments go to affiliated companies;

2. foreign subsidiaries benefit from invisible, that is free-of-charge, imports of R&D results from their parents or affiliates abroad.

The first condition is documented in Table 19. The table shows foreign- and Canadian-controlled shares of sales and R&D outlays before and after adjustment for R&D plus technology-related payments to non-residents: after adjustment the shares "even out."

Invisible transfer of technology

In order to verify that the second condition holds, namely that foreign subsidiaries benefit from invisible, free-of-charge imports, stronger assumptions must be made than with respect to R&D and technological payments.

The first assumption is that both parent/affiliate and subsidiary produce similar products using similar processes, and that R&D per-

---

9   See, for instance, Economic Council of Canada, *The Bottom Line*, Ottawa: 1983, pp. 40-42 and Statistics Canada, *Industrial Research and Development Statistics 1988*, Ottawa: October 1990, p. 77. But the as yet unpublished results of a 1992 federal survey appear to contradict all previous statistics that indicated a lower research intensity among foreign firms.

Table 19: Shares of sales, of R&D expenditures, and of R&D expenditures augmented (+) by technology-related payments to non-residents, 1975

Industry	Canadian-controlled			Foreign-controlled		
	% Sales	% R&D	% R&D+	% Sales	% R&D	% R&D+
Pulp and paper	56.4	67.2	52.7	43.6	32.8	47.3
Primary metals	82.9	86.0	78.9	17.1	14.0	21.1
Electrical products	34.4	59.2	53.2	65.6	40.8	46.8
Machinery	32.5	31.4	25.0	67.5	68.6	75.0
Chemicals	17.1	31.7	19.7	82.9	68.3	80.3

Source: Tables 5 and 10 in MOSST, *R&D in Canadian and Foreign-Controlled Manufacturing Firms*, Background Paper No. 9 (Ottawa, 1979).

formed by the parent is applicable to, and tends to flow to the Canadian subsidiary. The second assumption is that the amount by which the subsidiary benefits from the whole group's R&D is proportional to its share of the whole group's sales. In other words, in order to remain competitive, the subsidiary requires a research intensity similar to that of the parent. This is a very conservative assumption since the fruits of R&D within a multinational enterprise are most likely to be of the nature of public goods: any member firm of the multinational group is allowed access to it, no matter what its "taxes," i.e. transfer payments to the group are.

This leads to the following formula for the invisible imports of R&D by a foreign subsidiary in Canada:

RD (Invisible) = RD (Potential) – RD (Canadian) – RD (Payments)

where

RD (Potential) = RD (Group)/Sales(Group) times Sales (Subsidiary)

RD (Canadian) = the subsidiary's intra-and extramural R&D

RD (Payments) = the subsidiary's payments to non-residents for technology aquired.

Note that the formula is conservative, not only on account of the public good aspect, but because *technology*-related payments (the only ones available at industry level) exceed *R&D* payments by about one third. Also, the formula does not preclude a negative figure: it is possible that a foreign subsidiary *exports* invisibly.

Because of the importance of this technology transfer, a detailed example of the calculation of 1981 invisible R&D imports is given here—for the three-digit SIC industry 374, pharmaceuticals and medicines.

The example shows a step-by-step procedure which is determined by the formula given on the preceding page. *Potential R&D of foreign pharmaceutical subsidiaries in Canada* is calculated by first compiling data on as large a sample of such firms as can be obtained (table 20) and then "projecting" these data to a total population of foreign drug subsidiaries.

Table 20 shows the group (world-wide) sales and R&D of 12 multinationals, and the sales and R&D expenditures of their Canadian subsidiaries. The potential R&D of these 14 subsidiaries is estimated at $35.6 million in 1981. To estimate the potential R&D of *all* foreign drug subsidiaries a "projection to universe" must be made:

a) calculate a multiplier

K = Sales of all foreign subsidiaries/sales of sample
= 1,128 million/$623.2 million = 1.82.

b) assuming that R&D in universe is proportional to sales in the same way as in the sample,

total potential R&D = 1.82 times $35.6 = $64.8 million.

Next, the *Canadian-performed R&D* of all foreign drug subsidiaries must be estimated. Very likely a fairly accurate figure could be obtained from a special Statscan tabulation. In its absence the estimate of 1981 Canadian R&D is derived very simply. 1981 R&D intramural outlays in the pharmaceutical industry were $51 million. Using the 85 percent sales share held by foreign subsidiaries, we calculate (0.85 times 51) that about $43.4 million was performed by foreign subsidiaries.

Finally, *net payments to non-residents for technology acquired* (more than just R&D outlays) by the drug industry amounted to $7 million in

1981 (payments of $17 million, receipts of $10 million). Perhaps 85 percent, or $5.95 million, are the foreign subsidiaries' net remittances.

Table 20: Global Sales and R&D and Canadian Sales and National R&D of 14 American Pharmaceutical Companies in 1981 (in $ millions)

Firm	R&D Global	Sales Global	Sales Canadian	R&D Potential Canadian
Abbott Prod.	108.9	2313	46.76	2.202
Am. Home Prod.	111.2	4083	104.2	2.838
Baxter	58.6	1485	34.24*	1.351
Bristol-Myers	142.0	3492	35.90	1.460
Johnson & Johnson	260.1	5375	54.41	2.633
Lilly	238.0	2762	24.41	2.103
Merck	266.4	2922	84.29	7.685
Pfizer	174.5	3236	23.42	1.263
Shering Plough	106.3	1809	27.17	1.597
Searle	80.3	933	21.02	1.809
SKF (Or SKB)	215.7	2607	67.39	5.576
Sterling	66.8	1791	23.95	0.893
Syntex	60.5	696	32.90	2.860
Warner Lambert	106.6	3376	43.18	1.363
Total			**623.24**	**35.633**

*Interpolated.

Sources: Global R&D and sales, *Business Week,* June 20, 1983; Canadian sales via Consumer and Corporate Affairs, *Compulsory Licensing of Pharmaceuticals*, Ottawa: 1983, p. 8.

Using the overall formula, we can now estimate the *total invisible R&D imports* by foreign subsidiaries in the Canadian pharmaceutical industry to be:

= \$64.8 - \$43.4 - \$5.9 = \$15.5 million in 1981.

It is well known that the Canadian pharmaceutical industry, despite its high foreign ownership, is a heavy spender of company-financed R&D funds. Nevertheless, its invisible R&D imports amount to a quite considerable 30 percent of own-performed outlays. At the other end of the spectrum is the transportation equipment industry, SIC 32, which is largely foreign-owned and in which only the aircraft and parts industry, SIC 321, is domestically research-intensive. Estimated 1981 R&D invisible imports are \$669 million.[10] Somewhere in the middle lies the pulp and paper industry which is less than one-third foreign-owned and has a reasonable record of domestic research performance. Its invisible 1981 R&D imports are calculated to be \$21 million. Thus for the 32 SIC transportation group, imported R&D represented about double (\$669 million) the domestically performed R&D (\$296 million); for pulp and paper, SIC 271, it represented only about 26 percent.

Using the approach determined by the formula, a Ministry of State for Science and Technology background paper estimated 1974, 1975, and 1976 invisible R&D imported to Canada. Table 21 shows that in 1976 these invisible imports amounted to \$688 million, two-and-a half times higher than Canadian-located intra- and extramural R&D in these sectors. Thus, invisible R&D imports are likely to represent the most substantial component of all the R&D outlays that should be counted in to reach an adequate appraisal of an industry's technological intensity.

Statistics Canada applied the same methodology to later, 1983 data for transportation equipment (again) and chemical products industries. For transportation equipment the invisible imports amounted to a staggering \$827 million, seven times the domestic R&D outlays. For chemicals it was \$212 million, almost three times domestic outlays on research.

---

10 Calculations of imported invisible R&D for the transportation equipment and pulp and paper industries are given in Palda, *Technological Intensity*, Ottawa: Department of Regional Industrial Expansion, 1984.

Table 21: Invisible R&D to Canada, 1976

Industry Group	RD_{Pot} −	RD_{Cdn} −	RD_{Pay} =	Total RD_{Inv}
Mining	99.6	29.8	12.7	57.1
Petroleum	55.6	47.0	17.2	(8.6)
Food, Beverages, Tobacco	49.7	12.0	11.7	26.0
Machinery, Transp. Equipment	521.9	28.7	71.9	421.3
Electrical Products Utilities, Transportation	195.1	85.2	22.4	87.5
Rubber, Plastics, Chemicals	177.2	65.2	53.6	58.4
Metal Fabricating	45.6	3.9	12.6	29.1
Paper & Allied	34.2	9.3	7.7	17.2
	1,178.9	281.1	209.8	688.0

Source: MOSST, *Importation of Invisible Research and Development, 1974-1976*, Background Paper No. 3, Ottawa: July 1978.

If we agree with this type of analysis and even if we scale down the estimates of invisible transfers by one half, we would find Canada's GERD/GDP ratio reaching 2 percent.

Now it may be said that importance must be attached primarily to industrial research carried out in this country for reasons of employment and training of scientists, for not missing out on fundamental trends in technological development, and even to defend national sovereignty.[11] Such reasoning cannot be rejected out of hand, even if it is

---

11   National Advisory Board on Science and Technology, *Science and Technology, Innovation and National Prosperity,* Ottawa: April 1991, p. 22.

typically offered by self-interested industry, technology, and bureaucratic groups.

But there is an opposing viewpoint. Since the appearance of Adam Smith's *The Wealth of Nations* 200 years ago, the majority of economists believe that such wealth is best fostered when free market forces are allowed to seek out efficient, that is, lowest-cost solutions to economic challenges. The interest of the nation is taken to be synonymous with the interest of the buying public—the consumer in short. The consumer is best served when she has access to products or services benefiting from innovative modifications at the lowest price, no matter where such products or services spring from.

The consumer argument essentially states that it is immaterial whose lab the new product or process came from, as long as it is attractively priced. In the chain of cost components that add up to the final price, the R&D element is of some importance in innovative products. The lower it is, the less, obviously, the product's price to the buyer.

Government and university research on behalf of industry

Individual-industry research intensity is also typically understated by leaving out research tasks undertaken on behalf of industry, free of charge, in government and university laboratories. This writer has attempted to measure such costs elsewhere.[12] In a country in which the federal government performed $1.4 billion of research in its own facilities during 1989, accounting for 17 percent of GERD, some of this research must necessarily redound to the benefit of individual industries. On this it is useful to consult Box 4 which also shows that the federal government financed $629 million worth of university research (7.6 percent of GERD), with undoubtedly similar consequences. Unlike much government-performed research elsewhere, Canada's is mostly devoted to "civilian" technology, with a presumably closer connection to commercial outcomes.

---

12   Palda, "Technological Intensity: Concept and Measurement," *Research Policy*, 15 (1986), pp. 187-198.

But the government, usually so PR-conscious, is strangely silent on this issue. It does not in the least attempt to brag in comprehensive numbers about what approximate parts of the various ministries' budgets devoted to R&D seem to be close to the preoccupations of Canada's manufacturing and service industries.[13] The premier government research body, the National Research Council, overcautiously maintains that much of what it does cannot be assigned to individual SIC-type industries, since its research is often so advanced that the industries do not yet exist that may benefit from it.

It would require, therefore, a labour of Sisyphus to gather more than partial evidence on the pertinence of government-performed research to industrial interests. Table 22 contains a small illustration of the approach in which a tiny part of the annual disbursements of the Natural Sciences and Engineering Council to university researchers is assigned to individual industries.

The conclusion drawn here is that an industry's alleged deficiency in technological intensity needs to be carefully examined before any taxpayers' funds are committed to support increased research effort.

The Trade Balance Output Proxy of Innovativeness

Rightly or wrongly, balance-of-trade statistics are often used as the principal criterion of a country's competitiveness on the world markets. We need not elaborate here on the cyclical and stage-of-development pitfalls of this measure. We must, however, register strong misgivings about deploring an individual industrial sector's trade deficit without considering the total trade balance. Furthermore, it is particularly pernicious to reason on share-of-world trade grounds amidst strongly expanding overall world trade.

---

13   This is not so in the case of the one-client ministry of agriculture.

Table 22: Classification by 2-digit SIC industries of NSERC strategic grants to university researchers. December 1983, selected figures.

Total (covers biotechnology, communications & computers, energy, environment, toxicology, food & agriculture, oceans, open)	No. of grants	Amount
	497	$28.1 million
Energy[a]		
Mineral fuels	10	$715,303
Paper	4	124,182
Primary metals	3	178,560
Transportation equipment	3	131,168
Electrical products	13	672,931
Petroleum & coal	13	577,456
Chemicals	2	66,226
Construction	7	475,053
Electric power & gas	9	371,758
		3,313,087
Not classifiable	70	4.487 million
Total	134	7.8 million

[a]Three examples of classification are given below:

Industry	Project	Grant ($)
Mineral fuels	Fossil fuel potential of carboniferous pull-apart basins	122,796
Paper	Fluidized bed recovery of Kraft black liquor	31,000
Construction	Computer simulation and retrofit strategy for existing houses	47,620

Source: Natural Sciences and Engineering Research Council of Canada, *Information* (Dec. 2, 1983) Ottawa.

Table 23: Canada's Trade in High-Tech Products

Year	Imports	Ex-ports	Negative Balance		Trade Balance/ GDP (in %)	Ex-ports/ Im-ports
			Current Dollars (billions)	1981 Dollars (billions)		
1980	10.5	5.9	4.6	4.7	1.49	0.56
1981	12.9	7.4	5.5	5.5	1.54	0.58
1982	11.9	7.7	4.2	3.9	1.14	0.65
1983	13.5	8.4	5.1	4.6	1.27	0.62
1984	17.6	11.2	6.4	5.4	1.45	0.64
1985	18.4	12.1	6.3	5.0	1.35	0.65
1986	19.9	12.9	7.0	5.0	1.40	0.65
1987	20.7	13.6	7.2	5.7	1.31	0.65

Source: ISTC, *Science and Technology Economic Analysis Review*, and National Income Statistics.

Those who state that Canada is losing out in the innovativeness game, after showing us GERD/GDP ratios, point to the balance of trade in high-tech products. Consult Figure 6 to this effect. (The publication from which it is taken does not define hi-and the other techs, but it is likely to be OECD-2, as mentioned in Table 8.)

Table 23 covers the last 8 years of the high-tech trade shown in Figure 6. Exports, in current dollars, have more than doubled. The negative trade balance in 1981 constant dollars has not shown any tendency, and similarly the ratio of exports to imports and the deficit-to-GDP ratio. This table does two things that the doomsayers typically keep silent about: indexing to constant currency values and comparing to other important measures of the sector's or economy's performance.

Let us next turn to another trade indicator, to the balance of payments for technological *services*. Table 24 indicates, if anything, that by 1988 Canadian firms were not in deficit, and not very much prior to that as far as that trade is concerned. A more detailed examination by Hanel points out that telecommunications and office machines are chief contributors toward a positive balance.[14] (Hanel also has astute observations regarding patents taken out in Canada by Canadians and foreigners as output indicators of innovativeness. We shall return to this topic shortly when we discuss diffusion).

Finally, what about overall trade in manufactured products? Table 25 is one of the many possible indicators of trade performance, chosen because it does not rely on the share concept. Canada's performance over the years in manufactured exports does not seem to have been deficient compared to that of other industrialized countries.

Table 24: Foreign Payments Made or Received for Technological Services, Selected Years

Year	Payments		Receipts		Balance		
	R&D	Other	R&D	Other	R&D	Other	Total
	in millions of $						
1971	52	58	25	6	-27	-52	-79
1979	138	213	73	21	-65	-192	-257
1984	199	441	516	30	317	-411	-94
1988	343	507	830	60	487	-447	40

Source: Statistics Canada, *Industrial R%D Statistics 1988,* Ottawa: October 1990, Table 5.1.

---

14   Hanel, *op. cit.*

Table 25: Volume of Manufactured Goods Exports (1980=100) Selected Countries, Ranked by Increase, 1965-1989

	1965	1970	1975	1984	1986	1987	1989
Canada (3*)	27.2	57.2	70	105	108	117	122
U.S.A. (1)	31.7	45.5	65	123	130	138	142
Belgium-Lux (8)	32.4	53.8	63	73	106	113	112
France (8)	26.9	46.2	64	81	110	114	112
Germany, F.R. (4)	30.3	54.5	64	75	118	121	119
U.K. (6)	40.7	53.8	50	79	107	120	116
Japan (2)	20.7	42.0	66	95	126	139	138
Austria (10)	24.8	49.6	68	78	117	123	110
Sweden (5)	34.5	54.5	61	75	108	117	118
Switzerland (7)	38.6	57.9	59	75	120	120	114

*rank in 1989

Source: *U.N. Monthly Bulletin of Statistics*, Vol. 46, No. 3, 1991, p. 254.

Adoption/Diffusion as Part of Innovativeness

As we stressed earlier, the principal component of innovativeness is the firm's willingness and ability to adopt technology, whether from suppliers, licensers, or other providers. The mechanism of adoption/diffusion was explained in chapter 3 and its determinants briefly discussed; a lengthier enterprise-focused examination of adoption was offered at the end of chapter 4. But while we know quite a lot about individual adoption processes, we find it generally more difficult to speak of an economy's or industry's receptivity to technology than to comment on their creativity.

As usual, the extremes are easy to spot. All observers agree that Japan's progress to its present technological status was based on at least two decades of furious absorption of the foreign technical arts. Until recently, if not to this day, the technological balance of payments (transfer of patents, licensing agreements, provision of know-how etc.) of Japan has been negative, giving indirect evidence of technology absorption.[15] The other extreme, of conservative ossification of industry thirty to fifty years behind modern practice, is the Soviet Union and its former vassals. Despite the best efforts of the KGB's industrial espionage directorate and the existence of certain exceptional branches of the defence industry, and despite a GERD/GDP ratio almost certainly quite superior to that of the U.S., the heavy hand of party control proved an insuperable barrier to fast adoption.

We have already mentioned patents: they are, if of domestic origin, indirect indicators of homegrown innovativeness; if of foreign origin, potential harbingers of technology transfer. The extreme tedium of patent statistics analysis which draws some conclusions on this score is illustrated rather well in an oft-cited piece published in the MacDonald royal commission series.[16] The conclusions are highly interesting and encouraging:

> Canada, it seems, fares reasonably well in world terms as a recipient of new technology. This country grants the third highest number of patents to foreign nationals, after the United States and Britain... Canada receives a large flow of detailed information on new technologies, which has the potential for use under licence by Canadian industry. The fact that much of this information is not readily available to other countries can only provide Canada with an advantage as a potential user... Canada, overall, responds well to changes in active and stagnant technologies.[17]

---

15   *OECD Science and Technology Indicators No. 2*, Paris: 1986, pp. 55-57.

16   N. Ellis and D. White, "Canadian Technological Output in a World Context" in D.G. McFetridge (ed.), *Technological Change in Canadian Industry*, Toronto: University of Toronto Press, 1985, pp. 43-76.

17   *Ibid.*, pp. 72, 73, 74.

Against this optimistic conclusion there is a more pessimistic assessment by an earlier consensus report of the Economic Council of Canada:[18]

> Our general finding is that new technology diffuses slowly into Canada from other countries. It also diffuses slowly from firm to firm and from region to region within the country.

A more variegated assessment is given by McFetridge and Corvari.[19] They point out that diffusion and indigenous R&D are often complementary and that therefore firms undertaking R&D will be more receptive to the adoption of new technology: they will know where to look for it and know how to absorb it. Perhaps their most important observation is that the diffusion process is facilitated by the free movement of both goods and equity capital (for the latter, read multinationals) internationally. We have already dwelt at length on the invisible R&D imports which are the principal transfer of technology from abroad to Canada.

We recall here the evidence on purchases of machinery and equipment offered by Lucas (Table 13) and that gathered by Terleckyj on productivity increases among downstream purchasers (Box 2). This reminds us that an increased free flows of goods and services across borders, enhanced by the Free Trade Agreement with the United States (and soon perhaps with Mexico), will inevitably stimulate the rapid adoption of new technology by Canadian industry. What is perhaps more important is that it will offer all consumers a wider choice of technologically-efficient products and services.

The Competitiveness Issue

To be innovative, a firm, an industry, or an economy needs either to undertake some new product or new process creation, or to adopt

---

18 Economic Council of Canada, *The Bottom Line: Technology, Trade and Income Growth*, Ottawa: 1983, p. 63.

19 D.G. McFetridge and R.J. Corvari, "Technology Diffusion: A Survey of Canadian Evidence and Public Policy Issues," in McFetridge (ed.), *op. cit.*, pp. 177-231.

rapidly relevant new technologies, or both. Innovativeness, vague as its definition is, undoubtedly feeds into the yet vaguer state of competitiveness. The mechanism behind this relies on enhanced productivity: new products and processes, invented or adopted, increase the efficiency (lower the costs) of productive processes in mining, agriculture, manufacturing, services, and in households. Thus, Michael Porter tells us that the only meaningful concept of competitiveness at the national level is productivity.[20] He—and others—then go on to state that defining national competitiveness as achieving a trade surplus or balanced trade is inappropriate, for such do not guarantee a nation's standard of living. He concludes that to understand and explain competitiveness we must focus not on the economy as a whole but on specific industries and industry segments.

We have already looked at sectoral balance of trade figures and found them satisfactory within our context. It remains for us at least to glance at an economy-wide comparison, over time and between nations, of Canada's labour productivity in table 26.

As of 1986, Canada's productivity level (GDP per employed person) was still ahead of the ten other nations with which we like to compare ourselves, and behind the U.S. Its productivity grew more slowly over the three-and-a-half decades than that of its ten competitors, but this is only to be expected if we believe in the catch-up hypothesis represented in Box 5.[21] We can safely conclude on the basis of several Canadian studies—see the already cited works by Postner and Wesa as well as Bernstein, for instance—that Canada's overall productivity received a boost from new technology. Yet it would be hazardous to venture an opinion on how much of a boost and whether this was "enough."

---

20   Michael E. Porter, "The Competitive Advantage of Nations," *Harvard Business Review*, March-April 1990, 85; U.S Congressional Budget Office, *Using Federal R&D to Promote Commercial Innovation*, Washington: April 1988.

21   For a more thorough explanation see Moses Abramovitz, "Catching Up, Forging Ahead, and Falling Behind," *Journal of Economic History*, June 1986, pp. 385-406 and Richard R. Nelson, "Diffusion of Development: Post-World War II Convergence Among Advanced Industrial Nations," *American Economic Review*, May 1991, pp. 271-5.

Table 26: Real Gross Domestic Product Per Employed Person: Comparative Levels in the United States and Eleven Other Countries (United States=100.0)

	1950	1960	1970	1980	1986
United States	100.0	100.0	100.0	100.0	100.0
Eleven-country average	44.3	51.7	63.6	76.2	78.9
Canada	76.9	80.1	84.1	92.8	95.0
Japan	15.2	23.3	45.7	62.7	68.9
Belgium	46.9	50.3	62.2	79.7	81.3
Denmark	49.0[a]	53.5	60.1	66.6	68.8
France	36.9	46.1	61.9	80.2	84.3
West Germany	32.2	49.2	61.7	77.4	80.9
Italy	30.9	43.9	66.4	81.0	82.9
Netherlands	56.7	64.2	78.0	90.7	86.3
Norway	44.5	52.0	58.5	75.1	80.2
Sweden	44.0[a]	51.8	62.6	66.6	68.8
United Kingdom	53.8	54.2	57.9	65.8	70.4

Based on purchasing-power-parity exchange rates.
[a]Extrapolated by the author

Source: John W. Kendrick, "International Differences in Productivity Advance," *Managerial and Decision Economics*, Special Issue 1989, p. 14.

Of course competitiveness does not live by innovation alone,[22] although Porter, in his report on Canada's competitiveness, makes the

22   The glorious exception is the pharmaceutical industry where new products are the essence of continued survival. See Henry G. Grabowski, "An Analysis of U.S. International Competitiveness in Pharmaceuticals," *Managerial and Decision Economics*, Special Issue 1989, pp. 27-33.

point that "innovation—in its broadest sense—is the critical requirement for economic upgrading and increased prosperity."[23] And it could
well be that it is precisely the governments who wring their hands over
our alleged loss of competitiveness that are the institutions which are
most responsible for it.

To cite the famous Porter again:[24]

> What a new theory must explain is why a nation provides a
> favourable *home base* for companies that compete internation-
> ally. The home base is the nation in which the essential compet-
> itive advantages of the enterprise are created and sustained.[25]

Do Canadian governments provide the right environment for such
a home base? Here are three opinions and one newspaper story on this
issue.

From a letter by J. Laurent Thibault, president of the Canadian
Manufacturers' Association to Prime Minister Mulroney, dated September 25, 1990:

> Excessive government demands on financial markets have also
> resulted in a run-up of Canada's foreign indebtedness to 35
> percent of GDP or $230 billion in 1989. The resulting high
> interest rates needed to attract foreign money to finance the
> deficit means that Canadian industry is paying one of the high-
> est real costs of capital in the world. There is also mounting
> evidence that the overall burden of taxes and levies of all kinds
> on industry is now too high and is hurting our competitiveness.

From Peter Cook's "Trade Failures at Home and Abroad" in the
Globe and Mail, December 10, 1990, page B4:

> Canada's high-interest-rate, high-dollar policies have led to one
> of the sharpest deteriorations in competitiveness and in export

---

23   Michael E. Porter, *Canada at the Crossroads*, Business Council on National
      Issues and Government of Canada, October 1991, p. 73.

24   Ibid.

25   The importance of the home base is discussed at length and doubted in
      Richard Lipsey's majestic draft of *Economic Growth: Science and Technology
      and Institutional Change in a Global Economy*, Toronto: Canadian Institute for
      Advanced Research, May 1991, pp. 168-175.

market share of any industrial country. . . . Measured in U.S. dollars, unit labor costs in manufacturing went up 18 per cent in two years, 1988 and 1989, while in the United States they went down 9 per cent.

From the Economic Council of Canada's *Au Courant*, No. 3, 1991 article on Canadian productivity.

The Council economists also believe that a significant change in the current Canadian fiscal and monetary policy mix would have a positive impact on productivity and competitiveness. Tighter fiscal policies (both federal and provincial), to address budget deficit problems, would simultaneously allow interest rates and the exchange rate to fall and thus improve Canada's cost competitive position. . . . According to their analysis, Canada's productivity and cost performance can also be improved by policies that: facilitate and strengthen the ability to adjust to longer-term structural changes by increasing market flexibility; and increase domestic competition by removing national and interprovincial trade and investment barriers.

Finally, from a story in the *Financial Post*, June 15, 1991, pages 1 to 2, "The Last Straw—NDP Final Blow As Ontario Faces Slow Flight of Investment Capital":

Three years ago, the signing of the Canada-U.S. free trade agreement brought John Wood's ambition of making his appliance manufacturing company a North American player closer than ever.

To serve that market, Wood, president of W.C. Wood Co. Ltd., had already bought 140 acres near Guelph, Ont. He'd always assumed the expansion would be there. He grew up in Guelph and the company his father started in 1930 employed all its 700 workers there, making freezers, humidifiers, and range hoods.

But a study comparing the cost of expanding in Guelph to Ohio gave him a shock. Serviced industrial land in Ohio cost about U.S. $7,500 an acre compared to $100,000 around Guelph. Municipal taxes were lower by 60 percent, corporate income taxes by 12 percent and interest rates by 33 percent. The study concluded the C$ would have to be at US65¢ to make a new plant competitive in Canada. Last summer, his new freezer plant employing 100 opened in Ottawa, Ohio.

"Ontario in particular and Canada as a whole are losing all of the benefits that they gained under the free trade agreement through items which we do have control over," says Wood, citing the high C$, interest rates and especially escalating government spending.

Conclusions

We have shown that many of the well-known complaints about Canada's lack of innovativeness do not hold up to careful scrutiny. If, therefore, it ain't broke, to cite the popular maxim, why call for fixing it? And we come again to the opinion that if policies are to be invoked to increase innovativeness and even competitiveness, they should be of the broad framework, target-neutral kind that would have kept Mr. Wood's new freezer plant in Canada. And, indeed, in the first annual report to the United States President and Congress, the Competitiveness Policy Council identifies as the first line priority issues for its attention saving and investment (read Government deficits and tax policy) and education and training.[26]

---

26 *Building a Competitive America*, Washington: March 1, 1992. Technology comes next, but the Council does not perceive the problems there as lying within the government's purview.

Chapter 6

Government and Innovation in Canada

"That is, when the tax system, subsidy programs and R&D contracts are taken into account and when R&D contracts are assumed to be 50 percent subsidy, Canada provides a greater incentive to engage in R&D than does any of the other 11 countries for which these kind of data were available."

McFetridge and Warda[1]

Introduction

EVEN IF THE NEED HAD BEEN ESTABLISHED for an active industrial policy toward innovation in Canada—and the evidence of the preceding chapter certainly did not strongly point that way—another question should be raised before speaking to the kind of policy, if any, that would be appropriate. Has the support directed to the encouragement of innovation in Canada been clearly inadequate in level or ineffective in desired consequences? This chapter will attempt to cast light on the issue in a

---

1   Donald G. McFetridge and Jacek P. Warda, *Canadian R&D Incentives: Their Adequacy and Impact*, Toronto: Canadian Tax Foundation, 1983.

broad manner. We shall first discuss the scope and the kind of direct government support of innovation that has been available to industry in the past and is available now. This support is always undergoing change and no definitive listing of it can be offered. But it is, and has been considerable. We shall then present the normative case for the subsidization of innovative activities and the rules flowing therefrom, and ask whether there is any empirical evidence that such rules have been adhered to by the federal government. Next we will touch on some aspects of Canada's tax stimulants to R&D and discuss briefly international comparisons thereof. Finally, we shall look at some actual Canadian cases of government participation in innovative activities, from research in government labs to private sector subsidization.

The Extent of Past and Present Taxpayer Support—An Overview

The dollar amounts of federal and provincial expenditures—in house (intramural) as well as grants (direct subsidies) and contracts (partial subsidies of about 50 percent according to McFetridge and Warda)[2] have been outlined, for 1991, in Box 4. For ease of reference Table 27 reduces, simplifies, and adds 1984 statistics to the data provided in that box.

Three sorts of information emerge from the table: first, Canada's federal and provincial governments finance—and perform—a substantial share of the country's R&D efforts; second, their overall share in GERD has decreased from 43 percent to 36 percent as funders and from 25 percent to 18 percent as performers; third, this decrease in GERD share has been picked up by the business sector. The second and third developments are certainly in line with OECD expert advice and opinions.

The *size* of the taxpayer's contribution to the nation's research undertaking is, comparatively speaking, on a par with Japan's and Italy's with respect to the fraction of GERD/GDP financed by public

---

2   McFetridge and Warda, *op. cit.*

Table 27: Funders and Performers in Canada's GERD, 1984 and 1991 (in %)

Funders	1984	1991	
Federal Government	37%	29%	
Provincial Government	6	7	
Industry	37	42	
Universities	10	11	
Other	10	11	
	100%	100%	
	$6.0	$9.7 Billion	GERD
Performers			
Federal Government	22	15	
Provincial Government	3	3	
Industry	50	54	
Universities	24	26	
Other	1	2	
	100%	100%	

Source: ISTC, *Strategic Overview of S&T Activities in the Federal Government*, Ottawa: November 29, 1990.

funds. (Just as the Japanese and Italians do, Canada's taxpayers finance only a negligible fraction of GERD/GDP's ratio devoted to defence). This, too, rates a favourable assessment.

So much for the alleged insufficiency of the Canadian government's—or better, the Canadian taxpayers'—contribution to innovation outlays, an insufficiency proclaimed almost weekly from opposition benches and by media pundits. Is the distribution of this largesse deficient? Box 4 indicates that the federal and provincial governments together funded $533 million worth of business enterprise R&D expenditures in 1991, or 10.2 percent of it. This is almost certainly quite a wrong amount, a substantial underestimate; it does not take into account the tax advantages granted to R&D performers.

We can surmise that about one-half of the $465 million of federal financing consisted of government research contracts to take care of federal departments' research in-house and policy decision-making needs. The other half consisted of outright grants to industry. Since it was estimated by McFetridge and Warda that about half of the value of contracts represents pure subsidization of industry, we end up with roughly three-quarters of $465 million as industry subsidies, or about $349 million in 1991. But to this must be added the tax credits available to the private sector, and, of course, the total deductibility of both current expenditures on R&D and capital expenditures on equipment (though not on land and buildings).[3]

Table 28 shows a rough estimate of the size of tax credits and reasonably accurate figures on grants. By 1988 the estimated tax credits represented two-and-a-half times the size of the grants, and the two together exceeded $1 billion in private sector subsidization, or 13.5 percent of GERD, or 22.6 percent of business enterprise expenditures (BERD). Note that no contracts are involved in this calculation—the subsidization is "pure." This brings us to consider, piecemeal, some of the granting and tax incentive programs, past and present.

Direct Subsidy or Grant Programs

Table 29 reports on three taxpayer-financed subsidy programs for the ten years, 1980 to 1990. The oldest and most venerable of them (started in 1962) is run by the National Research Council. Called the Industrial Research Assistance Program (IRAP), it pays staff salary costs of selected research projects likely to initiate significant technological advance through commercial development and application in Canada. The selection criteria are the applicant's expertise in the field, and ability to effectively commercialize the research results. PILP, the Program for Industry/Laboratory Projects is designed to promote a more rapid transfer to industry of the results from NRC laboratories and other federal laboratories where there are important opportunities for Cana-

---

3   The immediate deductibility of capital expenditure for equipment is not available in many other countries, such as the U.S.A.

Table 28: Federal Grants Given and Investment Tax Credits Claimed with Respect to Industrial R&D ($Millions)

R&D Grants and Contributions	1978-79	79-80	80-81	81-82	82-83	83-84	84-85	85-86	86-87	87-88
(Fiscal Year)	76	103	114	154	168	221	252	280	280	303

Investment Tax Credits Claimed	1978	1979	1980	1981	1982	1983	1984	1985	1986	1987
(Calendar Year)	28	58	78	120	148	192	374	509	601	784

Sources: For grants Statistics Canada, *Federal Scientific Activities 1987-88 (1989)*, Cat. 88-204 and previous issues of it.

For investment tax credits, Statistics Canada, *Science Statistics*—April 1987, Cat. 88-001 to 1984 inclusive, 1985-6-7 figures from letter of Revenue Canada official, dated March 9, 1989.

Federal government grants are presented for financial years starting on April 1. The investment tax credits claimed are listed for calendar years.

dian industrial exploration. It has been administratively absorbed by IRAP since 1986.

The Enterprise Development Program, EDP, was created in 1977 and administered by the variously-named federal ministry of industry, nowadays and until further notice called Industry, Science, and Technology Canada. It was of assistance primarily to small and medium-sized businesses. One of its two goals was to support innovation; in that role it actually amalgamated and superseded a number of veteran granting programs run by Industry, Trade and Commerce. Grants were provided for new product or new process development projects which had the potential for profitable commercial exploitation. Up to 75 percent of the costs of approved projects could be contributed by the government to companies with sales of less than $10 million annually and up to 50 percent for larger companies. In 1980 the EDP's coverage was increased to provide special assistance to the electronics industry,

Table 29: The Three Major Federal R&D/Innovation Grant Programs, 1980 to 1990

		Thousands $		
Year	IRAP	EDP	IRDP	DIPP[d]
1980/81	20,691	47,000[c]		121,300
81/82	23,935	69,400[c]		151,600
82/83	31,511	65,800[c]		151,900
83/84	37,103	70,114	16,836[b]	169,200
84/85	40,635		46,887	130,700
85/86	37,247		51,006	203,192
86/87	54,817[a]		55,871	109,303
87/88	51,052[a]		35,233	135,158
88/89	62,543[a]			251,000
89/90	57,543[a]			300,909[e]

a. Combined with PILP
b. Started July 15/83
c. Budgeted, not actual
d. Estimates
e. Actual

Sources: For *IRAP*, National Research Council annual reports; for *EDP*, Tarasofsky, see footnote 10; for *IRDP and DIPP*, Dept. of Regional Industrial Expansion and ISTC annual reports and OECD, DSTI, *Financial Incentives for Innovation: Canada*, Paris, November 1984, and Dept. of Finance, *Canada Budget Estimates, Part 3, various years.*

and support for large-scale projects or company mergers that increased electronics production or research and development. Ottawa's Silicon Valley North was undoubtedly a prime beneficiary of these measures.

In 1982 the STEP (Support for Technologically Enhanced Productivity) program was introduced as a major extension of the special electronics fund (SEF) of the EDP. Given a budget of $20 million for 1982/83 it went beyond R&D to innovation adoption and diffusion support. At the same time, the scope of the Enterprise Development Program was enlarged to permit financing of market potential studies for innovative products and processes and other innovation-connected activities.

When the precursor to this volume came out, it carried on the above description by saying:

> In attempting to list in a comprehensive and concise manner the most important innovation assistance programs of DRIE (formerly Industry, Trade and Commerce, and Regional Economic Expansion) one is overwhelmed by the dynamic inventiveness of Ottawa's bureaucracy in this field. The phase has apparently been reached in which there is such a plethora of these programs that a consolidation and simplification of them is required. According to Bill C-165 (passed in June, 1983) a new, all-embracing program designated as the Industrial and Regional Development Program (IRDP) is being put into place. Any company anywhere in Canada will be eligible to apply for aid, whether it seeks a grant to develop a new product, or to obtain a loan guarantee to restructure an outmoded manufacturing operation. Chances are that by the time this book reaches its readers yet another reorganization of DRIE innovation assistance schemes will be in the offing.

Well, the IRDP program, which was only partly designed to support innovation, was actually cancelled only about five years later and then slowly wound down. The confusion accompanying its early days was analyzed in Box 3. By the beginning of the 1990s we are back to a plethora of programs again, in part described in Table 3. The most complete and annually updated listing of these programs can be found in a CCH publication.[4] If there are any direct successor programs to IRDP, they are to be found in the Atlantic Canada Opportunities Agency Action Programs and the Western Economic Diversification Programs.

---

4   David Horsley et al., *Industrial Assistance Programs in Canada,* Toronto: CCH Canadian Limited, various years. See also the colourful brochures issued by ISTC, such as the *Support for Technology and Development*, Ottawa: 1989, which lists ten federal and at least thirty joint federal-provincial technology subsidization schemes. Finally, ISCT issues yearly reports in which the individual programs are described in detail.

The largest program in the quiver of the ministry (and the federal government) is the Defense Industry Productivity Program, DIPP, which has a substantial R&D/innovation component. It is aimed at defence industries and their subcontractors and designed to increase their technological competence for *export* activities. The aeronautics and electronics sectors are the most important recipients of grants and loans. Those grants and loans cover a substantial percentage of eligible projects. "Defence" has to be taken with a grain of salt. Aircraft engine development, traditionally a voracious consumer of these subsidies, is not exclusively of military interest. If there is any trace of strategic trade policy to be found in federal subsidization, it is perhaps within DIPP with its export orientation.

The preceding footnote mentions joint federal-provincial subsidy programs. But there are, as well, of course, purely provincial programs that dig into taxpayers' pockets to support exciting, if not necessarily economically viable innovation schemes. A splendid example is the Quebec *Fonds de développement technologigue*. Modelled in part after Ontario's Technology Fund, it was set up following the 1989-1990 budget speech and endowed with $350 million. Two of its important subsidy programs are *les projets mobilisateurs* (catalyst projects) and small and medium business R&D projects.

The catalyst projects are said to have stimulated research projects close to $200 million, at the cost of about $50 million of subsidy, by May 1991.[5] The grants are destined to *precompetitive* research: partnership between several enterprises and university research institutes or departments that aim to provide a technological solution which constitutes real progress and leads to an influential market position of strategic importance to Quebec's economy.

The generosity as well as the complexity of the scheme—when subsidy and tax savings are combined—is apparent from Table 30. The object of the illustration is a manufacturing corporation whose taxable income exceeds $200,000. The first horizontal part of the table looks at the cost of the expenses incurred outside of the enterprise which are

---

5   Ministère des finances, *1991-1992 Budget*, Québec: May 2, 1991, Appendix A, Section 2.3.

Table 30: Financement d'un Projet Mobilisateur

Hypothèse B: Un projet mobilisateur de 10M$, dont 5,0M$ de R-D et 5,0M$ d'autres dépenses. Entreprise de transformation, revenu imposable>200 000$

		Partage du financement		
		GVT Québec	GVT Fédéral	Entre-prises
Dép. admissibles à la subvention du FDT	A. Subvention pour projets mobilisateurs (50% de 5,0M$)	2 500 000 $		
	B. Économie d'impôt—Québec	154 000 $		
	C. Économie d'impôt—Fédéral		646 000 $	
	D. Dépenses autres que R-D réellement assumées par les entreprises 5,0M$ – (A+B+C)			1 700 000 $
Dépenses de R-D	E. Crédit de 40% pour projets mobilisateurs—Québec (40% de 5,0M$)	2 000 000$		
	F. Économie d'impôt—Québec	308 000 $		
	G. Crédit à la R-D—Fédéral 20% de (5,0M$—E)		600 000 $	
	H. Économie d'impôt—Fédéral		620 160 $	
	I. Dépenses de R-D réellement assumées par les entreprises 5,0M$ – (E+F+G+H)			1 471 840 $
	Total	4 962 000 $	1 866 160 $	3 171 840 $

eligible for the grant. These would be mostly expenditures for R&D incurred *extramurally*. The fund (FDT) finances 50 percent of the $5 million, or $2.5 million. Tax savings offered by Quebec (GVT means gouvernement) amount to $154,000, and by the federal government $646,000. The firm, therefore, pays only 34 percent of the $5 million cost, or $1.7 million.

The second part of the table concerns (mostly intramural) R&D expenditure. Tax credits and deductions offered by both governments amount to $3,528,160, giving a net cost of in-house R&D to the enterprise of $1,471,840, or 29 percent. Thus the total cost of the putative $10 million project comes to 31.7 percent after subsidies.

In sum, these are impressive figures, even when compared to Canada's competitors.[6] Why is it, then, that the response of overall R&D spending is so anaemic, at least in the view of Canada's technology critics? A partial, perhaps even substantial answer may be the unimportance of industrial strategy when measured against the general receptivity-to-business climate. The question we just asked is, if anything, reinforced further by a brief appraisal of tax stimulants.

Tax Measures in Support of R&D

We have already indicated, with the help of Table 2, that Canada's tax régime is one of the most generous in the world with respect to *R&D* expenditures. We stress the R&D because we wish to make a distinction between industrial research, a component, and innovation outlays. Here we go into more detail, relying primarily on three sources of information: the already mentioned CCH Canadian compendium on industrial assistance programs in Canada, the Conference Board report on the international competitiveness of Canadian R&D tax incentives mentioned in Chapter 1, and a report prepared for Industry, Science, and Technology Canada.[7]

---

6   OECD, *Industrial Policies in OECD Countries, Annual Review*, Paris: 1990, part 4.

7   Deloitte & Touche, *A Comparison of Tax Incentives for Performing Research and Development in Canada and the United States*, Ottawa: May 1990.

For taxation purposes, the federal definition of R&D is accepted by the provinces. It states that "scientific research and experimental development is a systematic investigation or search carried out in a field of science or technology by means of experiment or analysis." It specifically excludes such activities as market research, sales promotions, research in the social sciences, routine data collection, and prospecting for minerals, oil and gas.

Federal tax support for R&D comes in two guises. The first are the *deductions* allowed. Firms can either write off immediately not only current expenditures on R&D incurred in or out of Canada, but also expenditures on R&D machinery and equipment, or they can choose to defer them to a future year. Ontario strengthens deductibility by offering a so-called super-allowance.

The second and, internationally speaking, exceptionally generous provision for support comes with *tax credits*. These can be applied by enterprises against their income tax payable and, nowadays in Canada for small businesses, can be obtained from the federal government as a cash refund if no tax is payable—such as when losses are incurred.

The tax credits are allowed on qualifying R&D expenditures, reduced by the amount of government grants and domestic contract payments, if any. For large corporations (roughly with taxable income over $200,000 a year) the federal tax credit is 20 percent of R&D expenditures incurred; 30 percent in the Maritimes and the Gaspé peninsula. The credit is increased to 35 percent for qualifying Canadian controlled private corporations (CCPCs), with an income of less than $200,000, up to $2 million of R&D expense—and 20 percent on any above that limit. The credit is, however, taxable as income the following year.

Further tax credits are obtainable under Nova Scotia and Quebec tax laws. For instance, in Quebec there is a 20 percent refundable tax credit available in respect of R&D wages, increased to 40 percent if the corporation is Canadian-controlled and has an equity of no more than $10 million and assets of less than $25 million. In addition, a 40 percent tax credit is available to corporations that contract for research with universities or engage in research consortia formed by the government.

It is therefore clear that in Canada it matters in which province the research is carried out. This is documented in Table 31 which shows that the cost of the research dollar can be as low as 40 cents, not counting direct subsidies, of course. An international comparison of tax and subsidy régimes will be undertaken in the next chapter.

Normatively Correct Subsidies to Innovativeness

At the start we should remind ourselves that while most of the literature speaks of R&D subsidization, we know that the economically interesting issue is the support to innovativeness: both the creation and the diffusion of new products and processes. The alleged market failure that may necessitate the remedy of subsidy occurs with respect to innovativeness, not with respect to R&D.

We have already discussed at length the rationales for public intervention under the three headings of market failure and in respect of strategic trade considerations in chapter 2. Strategic trade policy with respect to innovation is hard to discern in Canada and nobody really knows what to prescribe for it, despite recent advances in game theory. Similarly, nobody knows what is a critical level of risk, or a minimum efficient size for innovative behaviour so that a policy response can be triggered.

We are thus left in what is, in any case, the most discussed form of alleged failure, namely *inappropriability*, as a reason for public policy intervention. In other words, it may be that the inability to appropriate "sufficient" returns from innovation activity inhibits the firm from carrying out a "socially optimal" level of such an activity.

Inappropriability is not, in common with many economic concepts, easy to define in operational terms. Expenditures on basic research are most often an investment in inappropriable knowledge.[8] A percentage of patents originating in industry *i* and most likely to be used in other industries *j* is some indication of spillover. An ambitious U.S. survey of industry, asking questions such as "How effective are patents in your industry to prevent competitors to duplicate a new product?" and "What is the average time to duplicate a patented, major product

---

8   But not always. See the discussion and reference to Mowery and Rosenberg in Chapter 3.

Table 31: After-tax Costs of R&D Expenditures in Ontario and Quebec, 1990

For Small R&D Performers in Canada Eligible for the 35% Tax Credit Rate			For Large R&D Performers in Canada Eligible for the 20% Tax Credit Rate		
	Ontario	Quebec		Ontario	Quebec
R&D Expenditure	$1,000	$1,001[1]	**R&D Expenditure**	$1,000	$1,000[4]
Quebec R&D Wage Tax Credit (40% of $500)	–	(200)	Quebec R&D Wage Tax Credit (20% of $500)	–	(100)
Federal R&D Tax Credit (35% x $1,000) (35% x ($1,000-$200))	(350) –	– (280)	Federal R&D Tax Credit (20% x $1,000) (20% x ($1,000 - $100))	(200) –	– (180)
Tax Saving from Deduction (23%[2] of ($1,000-$350))	(150)	–	Tax Saving from Deduction (44%[5] of ($1,000-$200))	(352)	–
Quebec only			**Quebec only**		
Federal			Federal		
13% of ($1,000-$480)	–	(68)	29% of ($1,000-$280)	–	(209)
Quebec			Quebec		
3% of $1,000	–	(30)	6% of $1,000	–	(60)
Ontario only			**Ontario only**		
Tax Saving from Super Allowance	(34)[3]	–	Tax Saving from Super Allowance	(46)[6]	–
After-tax cost	**$466**	**$422**	**After-tax cost**	**$402**	**$451**

[1] Assumes that 50% of R&D expenditures is salary and wages.
[2] 23% is an estimated combined effective federal and provincial tax rate for CCPCs.
[3] Expenditures net of investment tax credits times percentage for incremental costs for small performers times the provincial tax rate [($1,000-350) x .525 x .10)]
[4] Assumes that 50% of R&D expenditures is salary and wages.
[5] 44% is an estimated combined effective federal and provincial tax rate for non-CCPCs.
[6] Expenditures net of investment tax credits times percentage for incremental costs for large performers times the provincial tax rate [($1,000 - 200) x .375 x .155)].

Source: Deloitte & Touche, *A Comparison of Tax Incentives for Performing Research and Development in Canada and the United States*, Ottawa: May 1990.

innovation?" provides numerical proxy variables for the level of inappropriability in various industries.[9]

But inappropriability can often be gauged in individual, specific cases of innovative products or processes—it is the equivalent of the consumer surplus that is generated by it. A graphical depiction of increase in that surplus was shown in Figure 8, panel (b) as the diagonally striped area $P_1P_2R_1R_2$. (Consumer surplus was defined as the difference between the unit price each customer would have been willing to pay for the product and the actual price, multiplied by the relevant quantities purchased). Reasonably simple ways of estimating that surplus include asking questions as to how much an existing product's price could be lowered due to a new lower-cost process—and so how many more units could be sold.[10]

The prescription, then, is to offer the potential innovator a subsidy which would cover the gap between the present value of his future innovation outlays and the present value of future receipts from the innovation, using his customary rate of discount. Provided, of course, that the subsidy shall not exceed the present value of the consumer surplus generated, less costs of granting the subsidy. A handy formula summarizes the prescription for S, the subsidy:

$$g - G = S \leq B - G - C$$

where g is the private cost of the innovation outlay, G its private benefit, B the economic benefits to society (consumer surplus) from the project and C the cost of the subsidy's delivery; all variables deemed in present values.[11]

An equivalent way of approaching the determination of optimal subsidy has been outlined by Usher.[12] The projects will be worthy and

9   Richard C. Levin et al., "Appropriating the Returns from Industrial R&D," *Brookings Papers on Economic Activity*, 1987, 3, pp. 783-831.

10   The Canadian authority on this subject is Abraham Tarasofsky's *The Subsidization of Innovation Projects by the Government of Canada*, Ottawa: The Economic Council of Canada, 1984.

11   Tarasofsky, *op. cit.*, p. 17.

12   Dan Usher, *The Benefits and Costs of "Firm-Specific" Grants: A Study of Five Federal Programs*, Queen's University, Dept. of Economics Paper, Jan. 1,

the subsidies necessary when the claimants and grantors can show that three *incremental* conditions are satisfied:

1. The project is incremental to the *firm*, that is, the applicant firm must document, and the grant administrator verify, that the firm's project will cost more than the present value of its expected private benefit.

In more mundane terms, the firm must make risk adjusted projections, properly discounted, of the project's future flows of costs and revenues, much as it would undertake for any other capital investment. A negative private return, a deficit, would imply that a subsidy is needed. At the same time this calculation would indicate the amount of the subsidy, which otherwise—while incremental—may be excessive.

2. The project is incremental to the *market* in addition to being incremental to the firm, that is, there is no other firm that could have undertaken the project profitably without a subsidy.

The fulfilment of this condition requires some investigation of the innovative, or at least R&D, activities of current or potential competitors.

3. The project is also incremental to the *economy*, that is, the benefits accruing to the economy as a consequence of the innovation must be sufficient to offset the subsidy granted and the cost of its delivery.

Practically this means that the social benefits of the grant must be calculated by the government agency and diminished by the direct cost of the subsidy and its direct costs of *both* application and administration. If the remainder is positive, the grant is incremental to the economy.

While the private and public benefits can, in the opinion of experts, be calculated reasonably easily, there is greater uncertainty surrounding C, the cost of transferring the funds from taxpayers to subsidies. What *is* known is that such costs are much higher than the general public is aware of. Tarasofsky estimates them as follows:[13]

1983.

13   *Op. cit.*, p. 13.

Cost of conceiving and administering the subsidy program	$0.10
Cost to firms of applying for subsidies	0.05
Cost to taxpayers of tax compliance	0.06
Deadweight loss resulting from tax (arbitrarily positioned within the range of various estimates)	0.60
Total per dollar of subsidy given	*$0.81*

The consensus among public finance economists as to the surprisingly high marginal deadweight costs occasioned by small tax increases is as yet little known. The concept of the deadweight tax loss cannot be explored here, but good references exist.[14]

The conditions enumerated previously seem logical, eminently reasonable, and what is more, practically attainable. Have they guided the federal innovation grant programs? Tarasofsky, who examined in detail the three programs listed as largest during the late '70s and early '80s (in table 29), replied "no" quite emphatically.[15]

With regard to DIPP Tarasofsky found it impossible to measure, on the basis of the information available to the program's administrators, the inappropriable benefits accruing to foreign or domestic markets.[16] In respect of IRAP (but not PILP, merged with IRAP since 1986, a program aimed at technology diffusion from federal labs to industry), Tarasofsky stated that "the information formally required from the applicants is utterly incapable of permitting a rational judgement as to whether the project warrants subsidization."[17] Finally, regarding the

---

14 The basic reference is Arnold C. Harberger, "Taxation, Resource Allocation and Welfare," in J. Due (ed.), *The Role of Direct and Indirect Taxes in the Federal Revenue System*, Princeton: Princeton University Press, 1964.

15 Tarasofsky, *op. cit.*

16 Since DIPP is heavily export oriented, Canadian taxpayer-financed innovation would presumably also increase foreign customer consumer surplus.

17 Tarasofsky, *op. cit.*, p. 59.

EDP grant program, Tarasofsky said that its most striking administrative shortcoming lay in the failure to recognize the relevance (and to provide for the projection) of the proposed projects' inappropriable benefits.[18] The assiduous reader may wish to check the most recent criteria used by ISTC and the National Research Council for grants under DIPP and IRAP-R programs in the CCH Canadian handbook.[19] Hanel and Palda have looked at the grant programs in a more aggregated and summary way.[20] Using proxies for appropriability they found that the IRDP grant-receiving industries did tend to have difficulties in appropriating returns for R&D through patent protection.

Thus far the appropriability question has focused on *grants*. Since *tax* incentives, almost by definition, apply to broad classes of corporate taxpayers, inappropriability plays hardly any role in them. And because, as was seen in Table 28, tax credits have become by far the largest part of R&D subsidization, the theoretically correct approach to the stimulation of research in Canada is increasingly in abeyance. But even this theoretically correct approach is now being questioned, as will be pointed out in the concluding chapter, and further conditions for subsidization are invoked.

We made the point that most discussion revolves around subsidies to R&D rather than to innovativeness. Clearly the rules with respect to *innovation* are not different from those with regard to R&D: one merely puts the accent on *all* expenditures, not just R&D, that go into the launching of a new process, such as new plant and distribution facilities, prior market research, marketing launch outlays, and so on. There are some subsidy programs in Canada that take partial account of such "R&D complementing" expenses. Among them are DIPP, the Atlantic Canada Opportunities Agency Action Program (ACOA), and the Federal-Quebec EDP program.

---

18   *Op. cit.*, p. 40.

19   *Op. cit.*, IRAP-R p. 68, DIPP, p. 175.

20   Petr Hanel and Kristian Palda, "Appropriability and Public Support of R&D in Canada," *Prometheus*, December 1992, pp. 204-226.

The tax system, as already mentioned, offers preferential treatment strictly to R&D expenditures. As far as this author could ascertain from discussions with ISTC officials, Canadian corporate taxation managers, American experts, German and Japanese embassy specialists, and from chartered accounting firms' internationally-oriented reports, tax régimes in these other three countries are no different—only R&D expenses "qualify." (On this point, for example, see the exchange of faxes between the writer and the Japanese Institute of Science and Technology in Appendix 6A). A difference between Canadian and foreign R&D tax regimes is pointed out in Box 8.

A Word on Risk, MES and Diffusion

As yet, there have been no hard and fast normative prescriptions devised to advise subsidizers about risk, MES (minimum efficient scale), and diffusion. But some useful observations can be made.

The words "risk" or "riskiness" appear as partial criteria for grant approval in several programs (IRAP-R, DIPP, Quebec Industrial Development Corporation's research and innovation activities). Circular 86-4R2 (August 29, 1988) of Revenue Canada Taxation states, however, that "the scientific or technological uncertainty, rather than the economic and financial risk is important in characterizing scientific research and experimental development—and hence eligible activities" (p.6). And of course, as we have seen, the tax law is not inclined to extend favourable treatment beyond strictly R&D to include other innovation-connected expenses.

Figure 13 shows a well known time pattern of profit flows accompanying the research, development and launching of a new product. As drawn, it is expected that the drain on corporate profits will be largest not during the R&D phase, but in the initial manufacturing and marketing launch stage. It is perhaps at point R that the risk of the venture may appear to be the heaviest to investors. Yet there is no tax provision to alleviate the burden of risk at this moment in the new product's life, no *tax* credit for at least a part of the new manufacturing costs or for part of the marketing investment that trains sales representatives and diffuses information to customers.

Figure 14 is a slightly different, but real-life, illustration of the investment burden accompanying a significant new-product launch-

Box 8: A Shortfall in Canada's R&D Tax Generosity

One area where the Canadian tax system is not competitive with other jurisdictions concerns the deductibility of scientific research expenditures carried on in a diversified group of corporations. Many situations arise where a division's scientific research expenditure benefits are significantly eroded because the activities are carried on in a separate corporation and that corporation is going through a period of low profitability. The research and development expenditures must be carried on in this division, due to provisions allowing corporations to deduct research and development expenditures only where it is "undertaken by or on behalf of the taxpayer, and, related to a business of the taxpayer." The only exception is where payments are made to a corporation exempt from tax under the provisions of paragraph 149(1)(j). This is often costly and difficult to administer.

The solution selected by most tax jurisdictions is to allow a commonly controlled group of corporations to file consolidated tax returns as is allowed in the United States or to allow the transfer of losses as done in the United Kingdom and Australia. This could be implemented for federal purposes only. Provincial taxes are calculated separately on the federal return albeit based on federal taxable income; but this can be accommodated easily.

If Canada is to remain competitive and promote scientific research expenditures, an ingredient to this process should include a consolidated return or transfer of losses within a controlled group.

Figure 13: Typical Sales and Profits of a New Product

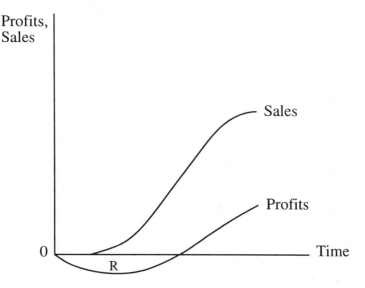

ing. The U.S. National Research Council proposes that net cash flows take more than 10 years before turning positive in a typical medium-size-and-range aircraft project, carried from the research stage to the delivery of 700 planes. Canada's income tax provisions allow tax credits strictly for R&D expenditures. What the preceding analysis suggests is that because the maximum risk of an innovative undertaking accompanies outlays that take place beyond the research stage, some provisions for innovation after R&D outlays should be made in the tax credit scheme.

As regards the MES, or minimum efficient size for a research undertaking (lab, engineering facility, software design team, etc.), there are two provisions currently and generously in place. Small firms which will never have a hope of undertaking research are amply served by federal and provincial ministry research facilities. Agriculture Canada spent over $325 million on R&D in 1990-91, Fisheries and Oceans around $135 million, Forestry Canada over $50 million. The National Research

Figure 14: Typical Cash Flow Curve for Large Transport Aircraft Program

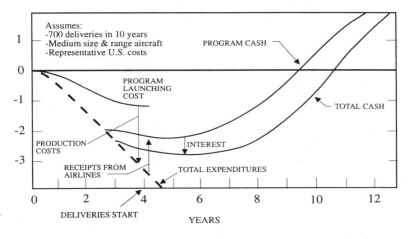

Source: U.S. Civil Aviation Manufacturing Industry Panel, et al., incl. National Research Council, *The Competitive Status of the U.S. Civil Aviation Manufacturing Industry*, Washington, D.C.: National Academy Press, 1985, p. 59, Fig. 2-9.

Council's primary mission is to help industry with research. For instance it spends in-house tens of millions of dollars on building research.

The second way by which alleged MES shortfalls are overcome is the federal tax credit (35 percent for CCCPs, only 20 percent for others, as already mentioned) and the various provincial tax provisions. Finally, many subsidy-granting programs have an in-built bias or open commitment to small and medium businesses.

Finally, diffusion of new technologies is by now probably in the forefront of both federal and provincial policies in support of innovation, as was already stated in chapter 2 in respect of the grants programs listed in table 3. First and foremost is the National Research Council, with the largest federal research budget—ahead of Agriculture Canada whose integrated task is both to invent and to diffuse—of which a considerable part is devoted to technology transfer:

> NRC's Industrial Research Assistance Program (IRAP) funds firms to adopt new technologies. In 1988, IRAP was a source of

technical assistance for roughly 5,000 firms. IRAP seeks the technology needed by a firm and transfers it by any appropriate means. The needed technology may be obtained from government laboratories, universities, technological institutes, other firms or foreign sources...

NRC, through the Canadian Institute for Scientific and Technological Information (CISTI), offers a renowned S&T information collection, dissemination and delivery service. CISTI responded to some 425,000 requests for material last year, of which about half were in support of industry...[21]

Other federal ministries, such as Energy Mines and Resources Canada, with its CANMET laboratories, are heavily engaged in transfer.[22] Industry Science and Technology Canada funds grant programs to support diffusion, such as the Strategic Technologies Program ($17 million in 1990-91), Technology Outreach Program ($18 million), Advanced Manufacturing Technology Program ($2 million), all listed in table 3, and the Technology Inflow Program for the acquisition of foreign technology. More detailed descriptions of transfer/diffusion programs are available in the already cited ISTC's *Strategic Overview of S&T Activities in the Federal Government 1989-90* and in National Research Council's annual reports. The preoccupation of the federal government with diffusion comes through via reports of the National Advisory Board on Science and Technology (NABST), such as the one of November 2, 1990.[23] That report quotes with admiration and envy the U.S. Federal Technology Transfer Act of 1986 which is designed to facilitate diffusion from government labs to the private sector.

Among the provinces, Ontario, for instance, is very active in the diffusion of technology. The Ontario Technology Fund in its annual 1989-1990 report states that the province spent between 1987/88 and 1989/90 $97 million on the establishment of so-called Centres of Excel-

---

21   ISTC, *Strategic Overview, op. cit.*, pp. 18-19.

22   Margot J. Wojciechowski, *Research and Development in the Mineral Sector,* Ottawa: EMR–CANMET Report CM 89-2E, August 1989.

23   *Revitalizing Science and Technology in the Government of Canada,* Section on technology transfer, pp. 38-45.

lence and committed $204 million to them.[24] Among their three objectives is the encouragement of the transfer and diffusion of technology to industry. The centres are located at universities, such as the Ontario Laser and Lightwave Research Centre at the University of Toronto, and the Ontario Centre for Materials Research, based at Queen's University.

We are thus led to the conclusion that the diffusion side of innovativeness receives very likely as generous a support from the Canadian taxpayer as the innovation side. A doubt, previously emitted with respect to innovation, applies as well to its diffusion. It is not clear why efficiently-managed firms *in the private sector* would not find it in their interest to inform themselves about ongoing technological advances or why they would not band together and insist that their associations provide them with this information. If they do not do this, then they either are not efficiently managed or they respond rationally to conditions in the environment which are not conducive to early adoption. In the first instance, should government really believe that it can improve managerial skills, at least in the private sector? In the second instance, should government intervene to alter the environment such as to increase competition through tariff reduction or to manipulate, for example, interest rates in order to ensure greater stability—and so long-term technological investment—in the construction sector? This type of argument does not, naturally, apply to areas of the economy (health, public administration, etc.) where innovation incentives do not operate.

How Effective has Government Support Been?

This study has looked at the formidable array of grant programs and tax stimulants to R&D, innovation, and diffusion at the federal and provincial levels. The next step is to ask of what effect they have been. Clearly, this is too ambitious a question and one that can only be answered piecemeal and incompletely. The answers can be broken down into two categories—the effects of tax stimulants and the effects of grants. They are not uniform, even within each category.

---

24   See also footnote 43 in Chapter 2.

Several authors have examined the impact of Canadian *tax policies*. In his 1985 contribution to the Macdonald royal commission, Bernstein came to the conclusion that tax incentives generally lead to a dollar-for-dollar increase in R&D expenditures. A year later Bernstein scaled his estimation down to an 80 cent increase per dollar of tax incentive.[25] In that same year Mansfield and Switzer estimated, on the basis of a survey sample of 110 American, 55 Canadian and 40 Swedish companies, that both tax allowances and tax credits increased R&D expenditure by about 30 to 40 cents for each dollar of tax abatement.[26] Writing a year later, Mansfield pointed out with reference to his previous study that:

> In all cases, the increased R&D expenditures due to tax incentives seemed to be substantially (and significantly in a statistical sense) less than the revenue lost by the government...moreover, in each country, there was substantial evidence that these tax incentives resulted in considerable redefinition of activities as R&D...of about 13 to 14 percent in both Canada and Sweden.[27]

What of the *grant* form of subsidies? We can ask, for instance, whether the R&D expenditures of private companies tend to increase (over what they would have been otherwise) as a result of receiving government subsidies.

In one of the best-known articles about R&D in Canada, Howe and McFetridge raised the question as to whether a direct government grant to a manufacturer acts as seed money for other R&D projects, or whether

---

25 Jeffrey I. Bernstein, "Research and Development, Patents, and Grant and Tax Policies in Canada," in D.G. McFetridge (ed.), *Technological Change in Canadian Industry*, Toronto: University of Toronto Press, 1985, pp. 1-41. See also Bernstein, "The Effect of Direct and Indirect Tax Incentives on Canadian Industrial R&D Expenditures," *Canadian Public Policy*, September 1986, pp. 438-48.

26 E. Mansfield, and L. Switzer, "The Effects of R&D Tax Credits and Allowances in Canada," *Research Policy*, 1985, 14, pp. 97-107.

27 Edwin Mansfield, "The R&D Tax Credit and Other Technology Policy Issues," *American Economic Review*, May 1986, 190-94. See also Mansfield and Switzer, "How Effective are Canada's Direct Tax Incentives?," *Canadian Public Policy*, 1985, 11, pp. 241-46.

it proves merely to be a transfer payment.[28] The authors had access to data assembled by the Department of Industry, Trade and Commerce which funded several incentive grants programs. Such programs took the form of cost-sharing agreements with the federal government paying one-half of the cost of R&D projects selected for support. While the firm must match out of its own resources the amount of public funds allocated to the *specific* project, there is no obligation imposed upon the total R&D budget of the enterprise. Consequently, the receipt of an incentive grant may increase, decrease or leave unchanged the total R&D outlays of the firm.

Keep in mind that the incentive grants programs are designed to support the realization of specific projects. The questions now raised go beyond this "narrow" scope. Quite apart, then, from the individual project focus, four types of outcomes are possible when the firm's total R&D budget is considered. The first, and the most "favourable," is when the grant increases the R&D expenditures of the recipient by more than the amount of the subsidy. This could happen when the returns to alternative R&D projects are interdependent. Let us write a simplified version of the Howe-McFetridge hypothesis as:

$$RD_i = a_0 + a_1 G_i + a_2 OV_i + e_i$$

where RD are total research and development expenditures *excluding* government subsidies, G the amount of government subsidies, OV are "other variables" such as sales, and e other influences, random in their aggregate effect, all for a given firm i. In this first and "best" outcome the coefficient a_1 would then be larger than 1.

If the grant is given to a project that would otherwise not have been undertaken and the recipient does not modify his other R&D projects, then a_1 would be approximately equal to 1. Should the recipient transfer the resources saved by the subsidy to other projects, his total R&D budget will be unchanged and a_1 will be between 0 and 1. Finally, the worst scenario of $a_1 = -1$ could happen if the grant is given to a project that would be undertaken in any case and the recipient decreases his total R&D outlays by the amount of the subsidy.

---

28 J.D. Howe and D.G. McFetridge, "The Determinants of R&D Expenditures," *Canadian Journal of Economics*, February 1976, pp. 57-71.

The authors, working with large samples of recipient firms observed over the period 1967-1971, found a statistically significant impact of government grants on total R&D expenditures only in the case of 104 subsidies to electrical firms. Holding the influence of other variables, such as the firm's size and its profitability constant, a dollar's worth of subsidy was observed to increase, on the average, the R&D budgets of domestic firms by about $1.30 and those of foreign subsidiaries by about 50 cents. In the case of 103 subsidies to chemical and 49 subsidies to machinery firms there was no statistically observable effect of the government subsidy on total R&D expenditures. By the same token, one can state that a_1 was *not* negative: the grants did not displace research expenditures that would otherwise have been undertaken.

In the authors' words the finding was "that despite the requirement that recipients of R&D incentive grants match public funds with their own, the overall commitment of resources to R&D by recipients of grants increased in only one of the three industries studied."

In 1984 Tarasofsky, using a regression model similar to that of Howe-McFetridge, rejected, at a high confidence level, the hypothesis that EDP subsidies had no impact upon the recipient firms' (96 subsidies in the electrical products industry, 88 in the machinery and equipment industry) discretionary R&D spending.[29] The regression coefficients indicated that each dollar of EDP subsidies paid induced, on average (between 1977 and 1980) increases in such spending of 47 cents and 63 cents in the electrical products and machinery and equipment industries, respectively. Yet EDP subsidies paid during this period were intended to cover half of the costs of the subsidized projects and the projects were intended to be fully incremental (meaning they would not have gone ahead at all without the subsidy). Since both of the estimated coefficients are less than one, it is clear that the latter intention has not been entirely fulfilled. It would therefore appear that substantial proportions of EDP subsidies seem to have been used by recipient firms as replacements for, rather than additions to, their own R&D spending.

Yet another aspect of the effectiveness of government stimuli to innovation concerns the contracting-out of government research to the

---

29   Tarasofsky, *op. cit.*, p. 79.

private sector in the hope of enhancing industrial R&D capability and producing "spin-offs." In a 1982 discussion paper by the Economic Council of Canada Supapol and McFetridge found that despite the excellent potential of the 1972 make-or-buy policy its net result has been that constant dollar industrial R&D contracting-out has actually fallen, while the less beneficial service sector contracting-out has risen.[30] As the regression analysis of the data revealed, one of the significant *negative* influences on the proportion of federal departmental budgets going to extramural industrial research was the percentage of the department's staff employed in intramural R&D. This would conform to economic theories of bureaucratic behaviour which would predict that effective resistance to the make-or-buy directives is likely to be more pronounced in departments where the scientific establishment is relatively large and influential.

We conclude this section with two observations. Governments in Canada, perhaps for excellent self-protective reasons, have not sponsored sufficient evaluation studies of taxpayer-financed innovativeness stimulants. What evidence there is does not lead to favourable conclusions. The evidence we cited is based mostly on statistical evaluations. Next we will examine some individual cases of government or rather, as we should continuously remind ourselves, of taxpayer-sponsored investments or policies in support of innovation.

The SRTC Fiasco and the Missing R&D Expenditure Bulge

In the fall of 1988 the *Globe and Mail* said that "the SRTC program resulted in what many government sources said was the largest tax fraud in the history of Canada."[31] It went on to state that, as of October 1988, 1,907 corporations took advantage of the scientific research tax credit program started at the outset of 1984; the total tax cost of the

---

30 A.B. Supapol and D.G. McFetridge, "An Analysis of the Federal Make-or-Buy Policy," Ottawa: Economic Council of Canada Discussion Paper No. 217, June 1982.

31 October 31, 1988, p. A4.

program was estimated at $3.5 billion, of which only $1.4 billion has been shown to have resulted in legitimate research and development. Revenue Canada has laid charges of income-tax evasion in 10 of the 61 cases under investigation. And yet the scientific and research communities have been strangely silent on this issue, well-investigated and reported by the media. While to this day that community does not miss a single opportunity to lament the demise of the ill-fated AVRO Arrow jet fighter for which the Diefenbaker government was blamed, it does not deplore the billions of the taxpayers' dollars wasted in the incredible SRTC boondoggle.

The SRTC story has many ingredients: a badly prepared legislative proposal, a minimal debate in the House of Commons, too slow a reaction of tax auditors, great accounting ingenuity and a mystery as to where all those tax refunds disappeared. It deserves a monograph of its own. The media could not do it justice, since by necessity they report piecemeal on unfolding stories. The best description and analytical insight is given in a little-known report by economist Donald McFetridge of Carleton University to the Auditor General, dated March 1986.[32] The following paragraphs use the McFetridge report as a source.

The intellectual origin of the tax credit was the notion that many innovation-oriented small companies needed cash up front for their R&D and could not wait for a refund at the end of the taxation year or claim a deduction against taxable income which they could not hope to attain in the early stages of their commercial life. This presupposed a perceived weakness in capital markets wherein it would be difficult to borrow against future tax benefits.

The remedial scheme concocted in the federal Department of Finance worked like this:

A firm performing (or wishing to perform) R&D sells an intrument to an investor priced at $100 and designates 50 percent of its value as a Scientific Research Tax Credit. On the same day the firm redeems the security for $55; the investor has now a $50 tax credit which he can apply

---

32   D.G. McFetridge, *On the Adequacy of the Information Provided to Parliament Regarding the Scientific Research Tax Credit: An Analysis of the Public Record, April 1983 - January 1984*, Ottawa, 1986.

against his own tax liability and $55. The R&D performing firm has $45 cash and must perform, within a year, $100 worth of R&D.

Note that the principal advantage to the firm F is that it gets $45 of cash without waiting for the tax credit reimbursement. This is presumably why it is willing to offer a $5 profit to investor I. But there is more to this.

Here is a date-related, extreme scenario.[33] On January 1, 1984 F issues, I buys, F redeems, I has $55 and $50 SRTC, F has $45 cash.[34] After January 1, 1985 I can redeem his $50 SRTC against his tax liability with the federal government. By June 30, 1985 F must deliver to Revenue Canada a financial statement for the year ending December 31, 1984. However, the beauty of the scheme is that firm F can take the $45 cash one-and-a-half years before its tax deadline.

The SRTC investor I can in no way be harmed by a dishonest R&D operator who may have run off 18 months before the tax deadline, for the law *did not make him liable* for the performance of R&D within the walls of firm F. He receives his tax credit no matter what.

The basic flaw, therefore, in this tax incentive was that it allowed investors to buy instruments without leaving any money in the firm issuing them. There was thus no incentive on the investors' part to distinguish between sound and bad R&D investments. This was rapidly discovered by certain leading accounting firms who designed the system of the so-called quick flips. Since they obtained advance rulings on the legality of such flips from Revenue Canada, they could sell their systems in perfect honesty. It is likely that the typical transaction returned 40 cents on the dollar to the R&D (and often "non-R&D") firm, after paying a 5 cent commission to the financial intermediary. But when you propose to develop submarines in landlocked Kitchener, Ontario and take the money south of the border, forty cents on the dollar is not bad as a net profit.[35]

---

33 *Op. cit., p. 16-17.*

34 This is a so-called quick flip.

35 "Known criminals such as Harold Arviv and con artists from the United States and Israel among other places, flocked to Canada to milk the SRTC program investigators say. Research ideas that were financed with tax

Perhaps the most brilliant R&D tax scheme under the SRTC label was revealed before a British Columbia provincial court in 1989. QIX Computer Corp. and QIX Facilities Corp. were alleged to have raised $13.9 million for projects to develop a complex computerized record-keeping system and a new process to preserve food longer by irradiation.[36] According to the *Globe and Mail* the promise of scientific research was never carried out. The plan for computer research called for $5 million in expenditure and qualified for $2.5 million in tax credits which Montreal-based Imasco bought for $2.25 million. Lawyers and brokerage fees claimed $250,000, leaving QIX with $2 million to undertake the $5 million project, the court was told by the prosecution.

In a letter dated March 9, 1989 to the writer Revenue Canada provided as an estimate $3.564 billion of SRTCs claimed. The program was in effect between November 29, 1983 and May 23, 1985, when it was shut down. McFetridge estimates a maximum potential revenue loss to the government of $2.6 billion. (Recall that tax credits are taxable as income the following year). He considers the minimum revenue loss to be $900 million, as Revenue Canada estimated that sum to be uncollectible (in 1986). He opines that the net tax revenue loss is under $2 billion. But the later (October 31, 1988) *Globe* article quotes Revenue Canada as saying that only $1.4 billion has been shown to have resulted in legitimate R&D. Thus McFetridge's estimates are superseded by later information.

Let us now try and pull all these figures together. We know, from the March 9, 1989 letter of Revenue Canada to the writer and from the October 31, 1988 *Globe and Mail* article that $3.5 billion worth of SRTCs were claimed over the short life of the program, concentrated essentially in the year 1984. The *Globe* article also mentions that only $1.4 billion "has been shown to have resulted in legitimate research and development."

---

credits ranged from guided missiles to exotic cattle, cruise ships to oil well fire snuffers and almost every conceivable variety of computer scheme." *Globe and Mail*, November 15, 1986, p. A4.

36   *Globe and Mail*, August 14, 1987, B3 and September 5, 1989, pp. B1 and 4.

Incredible as it sounds then, $2.1 billion was proving to be both a *fraud* and a *net loss* to the economy.

What of the $1.4 billion tax revenue given up by the federal government, or rather by the Canadian taxpayer? In some sense one cannot consider this a "loss" since the program was instituted by a democratically elected government. We can say, looking at the grant figures in Table 28 for the fiscal years 1983-4 and 1984-5 that it is not likely that the SRTC's were used *in lieu* of the regular research tax credits; they appear (the $1.4 billion) to have been, at least in part, *incrementally* used by the private sector to finance research.

If the entire $3.5 billion credit had been legitimately accommodated, $7 billion worth of R&D would have been performed. We know that in 1984—see Table 27—industry performed about $3 billion worth of research. But only $1.4 billion worth of SRTCs were "genuine," and so we would expect $2.8 billion of R&D to have resulted. Looking at the growth of intramurally performed R&D in industry we note[37]

1982	1983	1984	1985	1986
$2,489M	$2,585M	$2,994M	$3,610M	$3,949M
	4%	16%	12 %	11%

that while the rate of growth in such expenditures was impressive, the dollar figures between 1983 and 1985 rose by $1,025 million, nowhere near the $2.8 billion.

We simply do not know if we should have expected a massive bulge in Canada's business sector R&D expenditures over 1983-1985 or if $1.4 billion of SRTCs (= $2.8 billion of R&D) were used as a replacement for other existing tax alleviations. What we do know is that apart from some probing by the Auditor General no full scale enquiry into this fiasco was undertaken by the government, even though the calamity could have been laid entirely at the door of the Liberal government by the successor Conservative government.[38]

---

37   *Statistics Canada*, Service Bulletin, v. 14, No. 4

38   The program is launched in 1983, stopped very soon after the Conserva-

AECL-CANDU

There is no doubt whatever that the grand-daddy of high-technology ventures in Canada is the CANDU nuclear reactor; research and development activities associated with it persist and very likely still represent the single most expensive active hi-tech project around.[39] Since the largest proportion of the funds that went to the development of the heavy-water reactor was furnished by federal taxpayers, any—even a short—discussion of Canadian government support of research with a view to technological innovation must include CANDU.

In what follows we present a description and discussion of CANDU that was given in the 1984 precursor to this book, updated by the latest information. At that time, we deplored the lack of commercial success of Canada's nuclear venture, but we could not withhold our admiration of the technical performance.[40] In the last half-dozen years that technical prowess has come to be increasingly questioned.

The brief history will be followed by a description of costs incurred, benefits—economic and symbolic—secured, and future prospects. The guiding consideration is that CANDU may not have reached an economic breakeven point yet, nor even provided a net social benefit contribution.[41] As nuclear reactor programs go, this would of course, not be a startling phenomenon. For, to quote a well known article in this area, ". . . the commercial success of West German reactor development, as in other countries, has yet to be established."[42]

---

tives are elected in May, 1985.

39   This historical description draws heavily on Energy, Mines and Resources Canada, *Nuclear Industry Review-Problems and Prospects 1981-2000*, Ottawa: 1982 and on G. Bruce Coern, *Government Intervention in the Canadian Nuclear Industry*, Montreal: Institute for Research on Public Policy, 1980.

40   Except for one cautionary sentence that spoke of 1983 shutdowns which found Ontario Hydro to buy supplies from the United States.

41   George Lermer, *Atomic Energy Canada Limited*, Ottawa: Economic Council of Canada, 1987.

42   Otto Keck, "Government Policy and Technical Choice in the West German Reactor Programme." *Research Policy*, 9 (1980), p. 335.

The Atomic Energy of Canada Ltd. (AECL) was established as a federal Crown corporation in 1952 to carry on the research, development, and design activities connected with nuclear reactors that were undertaken under the wing of the National Research Council since the early days of the war.[43] (The first nuclear reactor to go critical outside the United States was Canadian, on September 5, 1945.) AECL soon turned its attention to electric power generation and in 1954 assembled a study team consisting of its own representatives and those of power utilities—most prominent among Ontario Hydro—and of manufacturers.

Because of wartime Canadian experience with heavy-water/natural uranium systems, and because the AECL wanted to avoid dependence on foreign sources of enrichment, and also because of attractive technological and engineering features the heavy-water/natural uranium design was chosen as a basis for the development of commercial power reactors. Since then, the evolution and development of CANDU (Canada-Deuterium-Uranium reactor system) has produced a technology which was particularly suited to Canada's industrial structure, resource base, and accumulated expertise.

The CANDU reactor design which emerged has several technological and economic advantages. First, the use of heavy water as a moderator and coolant combined with a design which maximizes efficiency allows the reactor to be fuelled with natural uranium; consequently, the efficiency with which uranium is converted to usable energy is high. Second, the use of pressure tubes, rather than a single large pressure vessel, as in light water reactors (LWRs), facilitates fuelling of the reactor while in operation, increasing plant availability. Finally, while making efficient use of natural uranium on a once-through basis, the CANDU fuel cycle can be modified fairly simply to embrace advanced fuel cycles, extending the usable energy obtainable from uranium resources (and eventually, thorium).

---

43 The research-to-revenue ratio of the Atomic Energy of Canada Limited Crown corporation is so high that even on "own-research" grounds only, the prefix "hi-tech" is more than justified.

When in 1955 the decision was reached to construct a nuclear power demonstration (NPD) reactor of the CANDU type, Canadian General Electric (CGE) was awarded the contract, as the government's stated policy was to create a private-sector capability for designing and building reactor systems. Private-sector companies, CGE and Deuterium Ltd., were also active in building Canada's first heavy-water plants. Midway through the construction project, in 1958, the federal government gave its approval for the design and development of a full-scale commercial 200 megawatt reactor. In June 1959 the government authorized AECL to begin construction of what was to be known as the Douglas Point (Ontario) reactor. This was done without waiting for the completion of the design and development phase, or for the NPD station (to be finished in 1962) to go into operation.

AECL—and not a private-sector manufacturer—was now to become the prime contractor and owner of Douglas Point, letting Ontario Hydro operate the plant and buy its power output. Later on, Ontario Hydro and public utilities in Quebec and New Brunswick assumed project management responsibility and finally ownership of reactors built in their respective provinces. AECL, however, took responsibility for reactor sales abroad and for two heavy-water plants in Nova Scotia (a much larger heavy-water facility is run by Ontario Hydro at the Bruce nuclear station). Private industry has been relegated to the role of manufacturing components and providing engineering services for reactors designed by AECL and built by provincial utilities.

The year 1959 represents a clear watershed between the original policy envisaging AECL's role as that of a primarily R&D organization providing support to private industry and a new policy of taking the predominant role in reactor design and construction activities. Doern lists the following reasons for this change which enlarged so drastically both the scope and size of the Crown corporation: the Diefenbaker government's attempt to redress the adverse image caused by the Arrow aircraft cancellation, the fear that any slowdown in reactor building may affect the Canadian uranium mining industry, the pressure from an impatient Ontario Hydro, and the alleged failure of Canadian

General Electric to control costs at the site of the demonstration reactor.[44]

The validity of the last point, the alleged inefficiency of private sector designers, was refuted by the successful and cost-effective performance of CGE in constructing a power reactor in Pakistan and an experimental reactor in Whiteshell, Manitoba. It is possible, however, to speculate that the parent U.S. General Electric Company may not have been enthusiastic about its Canadian subsidiary helping to perfect a reactor system rival to its own.

A rapid expansion of AECL's activity followed, fuelled mainly by domestic reactor sales and, to some extent, by export sales as well, and by the complementary heavy-water plant installation. Backing this impressive activity were federal appropriations to support nuclear energy research and development from the fiscal exercise 1947/48 to 1988/89 (AECL, as mentioned, took over from the National Research Council in 1952). Particularly to be noted is the last column of table 32 which converts the appropriation figures into 1986 dollars: the total sum of the Canadian federal taxpayer's investment in nuclear research alone to March 1989 is approximately $5.4 billion expressed in 1986 dollars. Figure 15 gives a graphic representation of these cumulative investments.

To this should be added other parliamentary investments which furnish the necessary support to the CANDU system: carrying charges on heavy-water plants and heavy-water inventories, losses on export sales of the reactors, and uranium industry subsidies. Mostly due to the very large surplus of heavy water, the federal government wrote off about $750 million worth of heavy-water *plant* investments in 1979/80. In the year ended March 31, 1989 AECL closed down its heavy water plants in Glace Bay and Hawkesbury (Nova Scotia) and in La Prade, Quebec, presumably because their production was surplus. The original cost of these plants was $803 million. In addition, one should take account of the deficits incurred in CANDU export sales. These are not easy to appraise.

---

44   Doern, *op. cit.*

Figure 15: Cumulative Constant 1986 Dollar Federal Expenditures in Support of Nuclear Energy R&D (in billions)

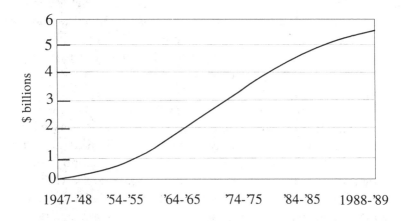

Source: Table 32.

While it is known that the Argentinian export contract incurred losses of well over $100 million, other aspects of the finances connected with export sales are less clear. It has been reported, for instance, that the preparation of the submission of the bid on the ill-fated Mexican venture and the maintenance of an AECL office in Mexico City cost around $50 million; this may overestimate the capacity of accountants to assess such outlays.[45] The stance of AECL's commercial mind-set in connection with this export venture is certainly disquieting, if it is the approach taken to sales abroad in general. It was reported that the organization's vice-president made the following statement: "As near as we can figure, if the Canadian taxpayer invested $1 in a nuclear sale, he would get in excess of $2 back in an eight- or nine-year period."[46] Suppose that the taxpayer would get $2.25 over a nine-year time span;

---

45   *Globe and Mail*, April 7, 1982, p. B14.

46   *Globe and Mail*, June 12, 1982, p. B1.

Table 32: Federal Government Expenditures Supporting Nuclear Energy Research and Development ($ millions)

Year	Current $[a]	1986 $[b]
1988-89	146.879	153.639
1987-88	153.700	156.837
1986-87	181.737	181.737
1985-86	176.612	172.641
1984-85	195.013	189.517
1983-84	184.446	177.012
1982-83	154.197	146.854
1981-82	145.715	148.235
1980-81	123.119	126.796
1979-80	114.654	112.076
1978-79	119.120	123.568
1977-78	128.490	138.758
1976-77	110.058	126.069
1975-76	104.922	127.333
1974-75	96.296	130.837
1973-74	95.163	143.102
1972-73	86.102	133.079
1971-72	88.768	142.714
1970-71	76.088	197.119
1969-70	74.400	201.678
1968-69	71.195	201.465
1967-68	69.000	201.616
1966-67	59.983	181.811
1965-66	54.267	172.220
1964-65	46.408	152.076

Table 32: (continued)

Year	Current \$[a]	1986 \$[b]
1963-64	32.063	107.585
62-63	37.832	129.374
61-62	34.633	120.187
60-61	38.828	135.172
59-60	31.156	109.840
58-59	30.798	110.748
57-58	30.394	110.812
56-57	32.845	122.320
55-56	32.886	126.998
54-55	20.506	79.6805
53-54	16.669	41.8423
52-53	17.704	69.7650
51-52	12.276	50.5147
50-51	7.327	33.5605
49-50	6.768	31.9837
48-49	5.890	28.8168
47-48	5.723	31.4272

Between 1947-48 and 1975-76 the contributions of the federal taxpayers went predominantly to finance AECL research and, in a minor way, to finance the Atomic Energy Control Board (never more than \$11.7 million in current dollars) and capital outlays for the NRU reactor and CRNL facilities (never more than \$12.5 million current).

Sources:
a. 1982-83 and later, Energy Probe *Report* 1991, before that Palda, *Industrial Innovation, op. cit.*, pp. 105-106.
b. Bank of Canada *Review*, March 1991, Table H4, GDP implicit deflator machinery and equipment, 1986 = 100.

the time of this statement (June 12, 1982) 1981 Canada Savings Bonds were paying 19.5 percent and the inflation rate stood at around 11.5 percent annually.

It is thus obvious that while the cash costs of the CANDU-related investments are relatively easy to assess, provided that they are put in perspective on a constant dollar basis, the opportunity costs, even if merely approximated by financing costs, are almost impossible to evaluate. The *total* federal taxpayer investment in nuclear energy can be estimated conservatively at close to $10 billion in 1986 dollars, of which between two-thirds and three-quarters went to broadly defined R&D. So much for the costs. What of the benefits to the Canadian economy yielded by this bold venture at the frontiers of science?

Here we are on even shakier grounds than with costs. It is surprising that a taxpayer investment which has now reached such a sum—an investment in fundamental and applied science, in manpower training and plant, and in national prestige—was never officially subjected to a credible overall cost-benefit analysis. What we know with some confidence is that in the mid-1970s the total direct employment in the Canadian nuclear industry (mining, R&D, engineering and design, manufacturing, construction, operations and maintenance, public administration) amounted to about 36,000 persons. We also know that well over one-half of Ontario Hydro's generating capacity is nuclear-based. We can reasonably conjecture that AECL's nuclear generation program provided excellent scientific and engineering training grounds for Canadians and a stimulus to a wide range of research, development, and other technological activities in Canada such as instrumentation, quality assurance, design engineering, and so on.

What we do *not* know is whether all this investment is making a profitable dent in the cost of electricity generation between, say 1965 and 1995; and whether we can indeed sustain the nuclear power manufacturing capacity—a jewel in the crown of our high-technology industries—to the year 2000.

It is almost impossible to make sensible predictions of the world market and even of the Canadian market for nuclear reactors. One senses and agrees with nuclear power advocates that global energy shortages and global warning will force a heavy recourse to nuclear electricity generation. But that state of affairs keeps getting delayed by

Three Mile Islands, Chernobyls and Mihamas and is always just around the corner.[47] On the other hand there are regions—France, Belgium, and Ontario are examples—where more than 50 percent of electricity is supplied by the atom.

The somewhat promising markets, apart from Ontario about which more further on, are located in the world's developing areas. (The industrially advanced countries are more or less locked in to their own suppliers or to those of their neighbours). Many of these areas are recipients of aid from the United States, UK, France, and Germany, or are dependent on political support from these nations. They are therefore fully open to political pressures from suppliers, generally desperate for business, based in these countries. Of course, AECL can and does enter the competition.

As is pointed out in its 1985/86 annual report (p. 2):

> Markets are most receptive to suppliers who demonstrate an innovative approach to financing, project management, technology transfer and, in some cases, long-term operation support.

Let us contemplate this innocent sentence. "An innovative approach to financing?" Why not an interest-free loan for the next 20 years to the customer, offered by the Canadian taxpayer? At the end of which the government of Canada forgives the defaulting country its debt.

"To project management and technology transfer?" We shall let the customer learn on the job and bring his engineers to Canada for training under some third-world-aid scheme. (It will pay off, to the customer, in plutonium and bomb-making capacity). "Long-term operation support?" If we are desperate enough, why not propose to build nuclear generators, run them in the customer's country, sell electricity locally and pay ourselves—in nontransferable dinars—for the project?[48]

What of Ontario, by far the predominant client of AECL and still its best prospective market? As this is written there is an internal debate within the socialist government of Ontario regarding Ontario Hydro's

---

47 *Time Magazine*, "Energy-Time to Choose," June 3, 1991, pp. 57-61.

48 We obviously make no reference to actual or contemplated sales of CANDU to India, Pakistan, Argentina or Turkey.

twenty-five year strategic plan. That plan states, in essence, that the existing system will fall short of supplying Ontario's electricity needs by 1995 and that a mix of energy-saving and energy-supplying actions must be taken. Ontario Hydro's demand/supply plan is the so-called "case 15" which requires the construction of 10 CANDU units at three stations, with a total capacity of 8.8 gigawatts (8,800 megawatts).[49] This plan clashes with the past NDP platform of no new nuclear capacity in Ontario. And it has been fully abandoned, by the spring of 1993, in view of both declining electricity demand and nuclear cost overruns.

Up to the mid '80s the heavy water Canadian reactor was one of the world's most proficient performers, as evidenced by the operating capacity indices in Figure 16, taken from AECL's annual report 1985-86. After that date such figures no longer appear in the annual reports.[50] What happened? From the annual reports of AECL, from newspaper clippings and from press releases of Energy Probe, a nuclear-sceptical lobby, it appears that CANDU reactors are aging much faster than expected. Of course, aging faster than expected is not confined to the CANDU design; it does, however, affect cruelly optimistic forecasts built into utilities' (such as Ontario Hydro's) plans.[51] In 1984, Ontario Hydro reported that the four 750 MW units of the Bruce A generating station of Lake Huron had an average capacity factor of 94 percent. In 1990 it was 47 percent.

Consider, in tandem, the 1985 performances in Figure 16 and those in Table 33, gleaned from an interim report of the Ontario parliamentary watchdog Energy Board of 1990. The actual 1989 and the forecast 1990 to 1993 performances of CANDU are markedly lower (by up to 50 percent) than those in 1985. Energy Probe maintains that Ontario Hydro's projections of an average 80 percent-of-capacity-attained per-

---

49   Ontario Hydro, *Providing the Balance of Power*, Toronto: presumably 1989.

50   But the New Brunswick Point Lepreau 600 MW nuclear facility continues to be in the very front world rank, with a life time capacity factor (actual relative to potential performance) of 93 percent at the end of 1990, over its life of eight years.

51   A Gloomy Year for Nuclear Plants," *Globe and Mail*, January 7, 1991. See also "Nervous about Nukes," *Time Magazine*, February 25, 1991, p. 59.

Figure 16: World Ranking—Nuclear Reactor Performance

1985 Capacity Factor (%)		Lifetime Capacity Factor (%)	
Japan Hamaoka-1	99.0	Canada Pickering-7	89.3
Canada Pt. Lepreau	97.4	West Germany Grohnde A-1	88.5
West Germany Grohnde A-1	96.0	Canada Bruce-3	88.0
Canada Bruce-1	94.9	Canada Bruce-4	86.1
South Korea CANDU Knu-3 (Wolsung)	94.3	West Germany Grafenrheinfeld	85.2
U.S.A. Salem-1	93.9	Canada Bruce-6	84.7
Canada Pickering-7	92.9	France Gravelines-C5	84.6
U.S.A. Oconee-1	92.8	Canada Bruce-1	84.6
West Germany Unterweser	92.0	Japan Takahama-3	84.4
West Germany Grafenrheinfeld	90.8	Canada Pt. Lepreau	84.2
U.S.A. Nine Mile Point-1	90.8	West Germany Gundremmingen B	84.1
France Tricastin-3	89.2	West Germany Stade-1	83.9
Canada Bruce-3	88.6	West Germany Gundremmingen C	83.7
Japan FukushimaII-1	88.1	Belgium Doel-3	83.6
		U.S.A. Susquehanna-2	82.4

Source: Atomic Energy Canada Limited Annual Report, 1985-86.

formance of CANDUs over the 40-years estimated life of a typical reactor should be scaled down, as indicated in Figure 17.[52]

Without going into the technical details, the chief problem seems to be faster than expected metal corrosion, particularly in the reactors' pressure tubes. The replacement of these tubes, the consequent shut-

Table 33: Nuclear Production and Capability Factors

	1989 Actual	1990	1991	1992	1993
Annual Production GWh					
Pickering NGS 1-4	10,489	9,503	11,118	11,561	14,468
Pickering NGS 5-8	15,014	13,863	15,347	13,959	14,013
Bruce NGS-A	13,467	14,855	17,069	16,281	14,681
Bruce NGS-B	26,291	24,328	25,579	24,279	25,259
Darlington	0	2,661	11,586	19,149	25,436
TOTAL	**65,261**	**65,210**	**80,699**	**85,229**	**93,857**
Capability Factor %					
Pickering NGS 1-4	59	54 (49)	63	65	81
Pickering NGS 5-8	84	77 (78)	85	77	78
Bruce NGS-A	55	62 (60)	67	60	57
Bruce NGS-B	93	84	86	81	85
Darlington NGS	0	70 (67)[1]	77 (76)[1]	81	84

Notes: [1]New units coming into service part way through the year.
( ) Update

Source: Ontario Energy Board, *Interim Report* HR19, 1990, p. 89.

---

52   Energy Probe, *Energy and the Environment: What the Ontario Government Can Do*, Toronto: Letter to the Ontario Premier, November 7, 1990.

Figure 17: New CANDU Reactor Performance Projections

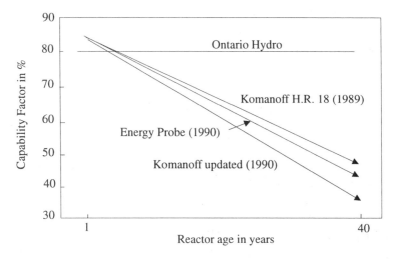

Source: Energy Probe, "Energy and the Environment: What the Ontario Government Can Do," Toronto: Letter to the Ontario Premier, November 7, 1990.

down and the necessity of generating or buying more expensive replacement electricity at the older Pickering and Bruce stations is estimated to have cost at least $1.3 billion since 1983.[53] Furthermore, technical difficulties in the construction of new generating stations simply do not seem to go away. Despite the billions of taxpayer dollars poured into CANDU research the technology still appears troublesome:

> Preliminary estimates made in 1973 indicated that the expected cost of Darlington was $2.074 billion (exclusive of commissioning and heavy water costs). It was then expected to be in-service in 1982. When the capital funds were finally released in 1978, the project's costs had increased by $2.9 billion (nominal dollars) to $5.0 billion, and the in-service dates changed to the period November 1985-February 1988. By 1981 a "definitive cost" estimate of $7.5 billion had been determined, with the in-service

---

53 Thomas Adams, "Commentary: Hardening of the Nuclear Arteries," *Globe and Mail*, November 13, 1990.

dates being further delayed to the May 1988-August 1990 period.

In 1983, the cost estimate was further revised to $11.4 billion (nominal) and the in-service period to May 1988-August 1992, in 1987 to $11.0 billion (in-service period: March 1989-February 1992), and in 1988 to $11.6 billion (in-service period: September 1989-February 1992). The latest revised estimate, which Hydro continues to designate as "preliminary," is $12.5 billion, a further increase of $939 million. Thus the cost, in nominal terms, has increased six times from the first estimate, and by 65 percent from the "definitive" estimate in 1981.

As indicated, each cost estimate change is associated with a change to the in-service dates. The most recent estimate is for the four units to come into service beginning February 1990 and extending over a period to December 1992. Thus the project has been delayed almost 10 years from its first estimated in-service date.[54]

(The *Globe and Mail* reports on January 7, 1991 that two of the four units have been started up).

Despite all of these discouraging developments, AECL can still rely on the taxpayer's bottomless purse. As its 1988-89 report states, a high priority is assigned to the design of the new CANDU 3, a 450-megawatt reactor destined for smaller utilities. The federal government (i.e., taxpayers) came through with a $44 million subsidy for pre-project costs, in addition to its annual R&D subsidy that amounted to $147 million, in that 1988-89 fiscal period.

There are perhaps three lessons that can be learned from Canada's experience with the production of nuclear reactors, lessons which should be pondered by hi-tech advocates of public largesse. The first is that technological accomplishments are easier to achieve than commercial ones and can, indeed, be bought by money. The second is that when government is a strong participant in a venture that must ultimately be justified on economic grounds, it will tend to discount market forces which will yet prove to be the stronger. Finally, when several industrial

---

54   Ontario Energy Board, *In the Matter of a Reference Respecting Ontario Hydro (H.R. 18)*, Toronto: August 31, 1989.

nations decide to back the same technological "winner," all of them may end up as losers in the ensuing rivalry.

Success in the Pharmaceutical Industry?

During the 1970s and much of the '80s an accusation was being hurled against the pharmaceutical industry in Canada: it did not perform sufficient local research and development. In the critics' eyes the reason for this was quite clear. The bulk of the industry was in the hands of multinational corporations who preferred to do research elsewhere.

It is quite true that the transnational presence looms large in the Canadian drug industry. It is on the order of 75-80 percent of manufacturing shipments. And it is also true that its average R&D/sales ratio of around 4-5 percent was only above one-half of that industry's ratios in some other industrially advanced countries.

The ratio, however, was not low on a more analytically-reasoned international comparison. Box 9 shows the regression model estimate and the ratio's "forecasts" that were used to determine the value of the R&D/sales ratio that should have been obtained in Canada between 1967 and 1977 had the Canadian industry's behaviour conformed to the international norm.[55] The average forecast of the ratio was minus 2 percent, the actual ratio turned out to be 3.7 percent.

But note the PRODPAT dummy variable whose coefficient essentially states that a strong patent protection increases the R&D/sales ratio, *ceteris paribus*. Since 1969 patent protection in Canada was substantially weakened by a federal law which imposed compulsory licensing on new patented drugs. About four years after a patent was registered, anybody had the right to import the medicine and sell it, provided of course that the drug was approved in Canada, and provided that the (compulsory) licensee paid a 4 percent royalty on his sales to the patenting firm.

Two consequences followed. The restriction on the intellectual property rights inherent in the temporary patents diminished substan-

---

55　The methodology was already explained in chapter 5 with regard to table 18.

Box 9: An Analytical Comparison of Cross-National Research Intensity

Between 1967 and 1977, every two years, data were obtained on the pharmaceutical industry in Belgium, France, Italy, Japan, Sweden, UK and USA. This time-series of cross-sections therefore had 7 times 6 or 42 observations, less one missing makes 41.

The regression relationship estimated was between R&D/Sales (for each country, for each year) and the following right-hand variables: PROFIT (profitability of a country's chemical industry measured by the rate of return on equity), INVCLIM (investment climate proxied by the ratio of retained earnings to total earnings in the chemical industry), CONCEN (four-firm concentration ratio in the drug industry) to approximate industrial structure, FOROWN (share of domestic sales supplied by the foreign-owned companies), PRODPAT (measuring strength of patent protection).

The result:

$$RD/S = 0.03 - 0.02 \text{ PROFIT} - 0.03 \text{ INVCLIM} + 0.003 \text{ CONCEN}^b$$
$$- 0.001 \text{ FOROWN}^b + 0.02 \text{ PRODPAT}^a$$

$$R^2 = 0.77$$

a and b designate statistical significance at 5 and 1 percent levels, respectively.

The forecasts and actuals of R&D intensity (% production value) for the Canadian pharmaceutical industry:

Year	Forecast	Actual Value
1967	-1.7	3.5
1969	-1.0	3.8
1971	-2.1	3.3
1973	-2.1	4.0
1975	-3.5	4.4
1977	-1.7	3.3

Source: Palda and B. Pazderka, "International Comparisons of R&D Effort: The Case of the Pharmaceutical Industry," *Research Policy*, 11 (1982), pp. 247-59.

tially the appropriability of the fruits of innovative behaviour. It aroused the ire of the multinationals who saw their patents threatened by Canada's example in their other markets. It naturally decreased their propensity to invest in research on Canadian soil. Research is the lifeline of this industry. A new chemical entity (NCE) or molecule costs about $150 million to bring to market, and so the inability to recover its costs because of imposed regulatory constraints touches the innermost interest of new-drug originators.[56]

The second consequence was the apparent lowering of prices of patented medicines in comparison to those prevailing in the United States.[57] This was a natural development, since the temporary patent monopoly on new drugs was breached by compulsory licensing. As a consequence the consumer interest (but above all the taxpayer interest because of the universal health care system in Canada) was enhanced. In the longer run, however, drug innovation probably slowed due to increased inappropriability; and thus consumer interest was actually harmed.

By the mid '80s pressure was building on the government to restore more effective patent protection to the industry. The task force on biotechnology in its 1981 report to MOSST alleged that the compulsory licensing provision has "a devastating effect upon investment in pharmaceutical research and development in Canada. At a time when other countries are strengthening their patent systems to ensure the patentee a fair return on investment and thereby stimulate industrial development, this provision in Canadian patent law has essentially eliminated all new pharmaceutical development."[58]

---

56 Henry G. Grabowski, "An Analysis of US International Competitiveness in Pharmaceuticals," *Managerial and Decision Economics*, Special Issue, 1989, pp. 27-33.

57 Harry Eastman, *The Report of the Commission of Inquiry on the Pharmaceutical Industry*, Ottawa: Canadian Government Publishing Centre, 1985, and especially its background studies published later on separately.

58 *Biotechnology: A Development Plan for Canada*, Report of the Task Force to the Minister of State for Science and Technology, Ottawa: February 1981.

The closing down of one of the largest pharmaceutical research laboratories in Canada, Ayerst of Montreal (subsidiary of American Home Products) has drawn the attention of the public and of the Quebec and federal governments to a possible choice they have to make between the interests of producers and the interests of consumers. The Department of Consumer and Corporate Affairs put out a discussion paper concerning draft legislation that would attenuate compulsory licensing and, at the same time, make such attenuation depend upon increased R&D.[59]

A furious debate developed then about Bill C-22 which the Conservative government submitted to parliament in 1986.[60] The bill was passed only in late 1987. Its main provision was to offer a 7 to 10 year protection to newly patented products from generic (i.e. non-brand) competition. The patent-holding manufacturers (all of them members of PMAC, the Pharmaceutical Manufacturers' Association of Canada) promised in turn to increase their research/sales ratio of about 5.9 percent in 1986 to 10 percent in 1996, or by about $1.4 billion to a total of $3 billion over the decade. This presumably would add 3,000 "hi-tech" jobs to the economy. As of October, 1988 PMAC members have published expansion plans totalling $1.1 billion by 1996.[61]

And, *mirabili visu*, for once an economic prediction made was actually fulfilled. Table 34 is based on annual reports of the Patented Medicines Price Review Board (PMPRB), a regulatory body established to monitor price changes and R&D performance. The most notable features of the table are the much higher than predicted R&D/sales ratio of PMAC patentees and the increasing percentage of research effort undertaken in the basic category. The board is also pleased to report, in June 1992, that between January 1988, when it was set up, and the end of 1991, the price index of patented pharmaceuticals has run below the consumer price index. Before then it rose faster.

---

59   Consumer and Corporate Affairs Canada, *Compulsory Licensing of Pharmaceuticals*, Ottawa, May, 1983.

60   Harvie André, *Notes for Opening Remarks to Legislative Committee on Bill C-22*, Ottawa: Consumer and Corporate Affairs Canada, 1986.

61   *Financial Post*, Special Report on Pharmaceuticals, October 31, 1988.

Table 34: R&D Statistics in Canadian Pharmaceuticals, 1988-1991 (All Patents PMAC)

Year	Sales[a] ($M)	R&D ($M)	All Patent- ees R&D/S (%)	PMAC Patent- ees[b] R&D/S (%)	Basic R&D ($M)	Basic as % of total	Extra- mural[c] ($M)
1988	2,178	165.7	6.1	6.5	30.3	19.1	n/a
1989	2,973	244.8	8.2	8.1	53.5	21.9	53.8
1990	3,299	305.5	9.3	9.2	78.4	24.9	69.4
1991	3,894	376.4	9.7	9.6	94.2	25.0	84.1

[a]63 to 66 pharmaceutical product patentees reported. They appear to account for all pharma R&D in Canada
[b]PMAC—Pharmaceutical Manufacturers' Association of Canada.
[c]Research paid for by firms but carried out in hospitals and universities.

Note: Almost all of patentees' R&D (96%) is self-funded.

Source: Annual Reports of the Patented Medicine Prices Review Board, Ottawa, 1989-1991.

Is this the case, as in the Czech proverb, of the wolf having eaten and the goat remaining intact? Only time will tell whether a good compromise has been achieved between the interests of the citizen as a consumer and as a taxpayer. What can be asserted is that the Conservative government's stance on this intellectual property issue was courageous in the face of powerful counterinterests, ranging from retired peoples' associations to Canadian superpatriots. Indeed, it was a preview of its fight over the Canada-U.S. free trade agreement.

We have now analyzed a federal hi-tech investment (in CANDU) and two federal government policies, one tax-, the other patent-related, one a horrible failure, one an apparent success. The fourth case is that of an innovative product, developed in the federal government research establishments.

CANOLA and Agriculture Canada

As we have seen, the outcome of the grandest of Canada's R&D mega-projects, the CANDU reactor, while originally a technical success, is yet to attain commercial viability. This does not necessarily mean that taxpayer-supported research ventures cannot result in substantial economic returns. One such instance of putting public funds to excellent use is the research carried out for many years—one is almost tempted to say from time immemorial—by the federal and provincial departments of agriculture. A good example here is the case of rapeseed which, when processed, yields both edible oils and protein meal for animal feed.

Far from the din of high-technology trumpets Agriculture Canada and university-based scientists in the Western provinces, in close cooperation with farm associations, succeeded in developing high-yielding varieties of this plant and "bred out" certain noxious acids first present. (To emphasize the qualitative improvement, rapeseed was rechristened "Canola.")

Rapeseed was first commercially grown in Western Canada in 1943. Starting in earnest in the early '60s, rapeseed research and breeding was successfully concluded in the mid-1970s, though "maintenance" work is likely to continue. The bulk of the expenditures was incurred between fiscal 1963/64 and 1974/75 and amounted to approximately $12 million in 1961 dollars. These outlays include activities devoted to both yield improvement and acid removal through breeding. The social returns could only be measured against the higher yields achieved by rapeseed, since the measurement of quality impact (digestibility due to lower acid content) proved too difficult. Thus they are probably understated. The returns were estimated to be over $86 million expressed in 1961 dollars.[62] Accordingly, the internal social rate of return was calculated to be 101 percent. In other words, looking over the 11 years from 1963/64 to 1974/75 the money invested in rapeseed development could have been borrowed at a 101 percent rate of interest without incurring a loss.

---

62   Joseph G. Nagy and N. Hartley Furtan, "Economic Costs and Returns from Crop Development Research: The Case of Rapeseed Breeding in Canada," *Canadian Journal of Agricultural Economics*, 1978, No. 1, pp. 1-14. See also John C. Hughson, "Canola—An Oilseed with a Future," *Canada Commerce*, December 1981, pp. 1-3.

Another interesting aspect of the Canola case was the calculation of the distribution of the social benefits between producers (the growers) and consumers (really commercial buyers). The buyers benefited from a lower price as the supply of the oilseed expanded. The growers, despite a falling price, expanded their revenues and profits. It was estimated that of the total social gain of $86 million, the consumers received about 53 percent and the suppliers about 47 percent.

The 101 percent internal rate is obviously a very attractive return and, in order to keep matters in perspective, should be set against some cross section of federal agricultural research projects. Not all of those can be successful, technically or commercially, given the uncertainties accompanying innovative endeavours. In the Canola case, however, there have been further attractive developments that are typical of the Canadian agricultural sector as a whole: oil exports increased from 43,000 tonnes worth $23 million in 1976 to 193,000 tonnes in 1980 worth $119 million. Oilcake and meal exports, during the same period went from 52,000 to 207,000 tonnes or from $6 million to $37 million.

Without knowing the detailed specifics of the case one may nevertheless speculate that two important phenomena accounted for much of the success of the Canola project. On the supply side, the technical advances were most likely of a step-by-step, established plant breeding technology kind, rather than fundamental breakthroughs. On the demand side, Agriculture Canada and university-attached agroconomists are well known for their close, continuous contact with the farmers, their demand constituency. It is possible to conclude that this careful attention to market and realistic technological ambitions were the key elements present.

Novatel

Wasting taxpayer's resources on R&D support or on high-technology fantasies is not a prerogative of federal or left-leaning governments. The short and brutish life of Alberta's Novatel is a case in point. It was well documented in the daily press, but despite that, this writer was not able to get hold of precise information concerning the corporation's R&D outlays between its founding in 1983 and its near-demise in 1991, and Alberta's government support for that research effort. Nevertheless, everybody is agreed that Novatel was a hi-tech investment choice in

Alberta's industrial policy of deliberate diversification away from oil and agriculture. It thus "qualifies" as our choice of case illustration of governmental innovation policies.

Novatel Communications Ltd., Canada's only manufacturer of cellular telephones was created in 1983 as a partnership between the Edmonton-based Alberta Government Telephones and Nova Corp. of Calgary.[63] In January 1989, Nova sold its half of the company for $60 million to Telus Corp., the recent name assumed by Alberta Government Telephones, AGT.

Novatel's two main sources of revenue are its "systems" business which makes switches, radio frequency base-stations and other equipment used by cellular phone companies and making "subscriber equipment," including car phones and hand-held cellular phones. Given the fast growing cellular telephone market in North America, Novatel since its inception has always been regarded as a star in the Telus Corp. constellation and was also expected by the Conservative provincial government to diversify the local economy of Alberta. The 1989 annual report of Telus Corp. stated that Novatel was the world's second largest seller of cellular phones during that year. It also said that Novatel was the top seller in Canada with 25 percent of the market and that in the U.S. it accounted for 21 percent of all sales. However, the Calgary-based company never posted an annual profit, even though it mushroomed from just three persons in 1983 to a staff of about 1,400 by the end of 1990.

The Alberta government sought a partner for the displaced Nova Corp. investment in Novatel. That company apparently required huge R&D investments to survive the growing global competition. The cellular phone industry includes such industry giants as Motorola and NEC Corp. of Japan. As well, an active partner with a strong international base was needed to expand Novatel's international market opportunities. Even the federal government's assistance to Novatel to the tune of $5 million sanctioned in early 1990 was conditional on the

---

63 Sources: *Montreal Gazette*: September 1990 to December 1990, *Calgary Herald*, November - December, 1990, *Financial Post*, September 1990 and, above all, the *Globe and Mail*, July 1989 to March 1991 and especially, May 28 and May 30, 1991.

company seeking an industry partner who would eventually acquire a substantial interest in its affairs in the next 18 months.

Robert Bosch GmbH, the West German giant in automotive electrical components, was lured into considering being such an industry partner to Novatel. By July 1990 Telus, the former AGT, was close to announcing the sale of half of Novatel to Bosch at a price rumoured to be $111 million.

Meanwhile, in the latter half of 1990, Telus Corp., the government-owned parent company of Novatel, went public, announcing the biggest ever equity issue in Canadian history. The Alberta government sought to raise $951 million by selling a 60 percent stake in Telus to the public. The province sold $816 million worth of shares to Albertans under an investment plan and an additional $35 million worth to Canadians inside and outside Alberta. The equity offering has become so popular that the province could only sell a maximum of 900 shares to each of the 140,000 Albertans who had applied.

It was during this privatization period that Novatel slipped from being a blue-chip company into an expensive millstone around the neck of the government, posing a massive economic and political liability. The cellular handset industry proved to be highly competitive, requiring large R&D expenditures and suffering from low profit margins due to imports from the Pacific Rim.

Both the forecasting and the accounting of Novatel officials went woefully wrong. Novatel executives had substantially overestimated the company's profit potential for the remaining five months of 1990. The shareholders who were promised a profit of $16.9 million for the second half of 1990, were apprised of an estimated loss of $4 million for the same period.

An expected profit of $3.6 million for 1990 turned into a $204 million loss.

The prospective buyer of 50 percent of Novatel, Bosch, backed out of the deal. Telus washed its hands of its own half. On January 1, 1991, it exercised an option to sell Novatel to the Alberta government for $159 million. That government announced later that it extended a $525 million line of credit to cover the cost of Novatel's operations. The opposition estimated that the government's losses were approaching $900 million.

The representatives of Alberta taxpayers did not, of course, confine their industrial diversification policy of picking winners to Novatel only. Some of their other ventures have been summarized in Figure 18, taken from the *Globe and Mail* of May 28, 1991. In closing, it is worthwhile to cite two sentences from that *Globe* article:

> Just last year (1990) Novatel was extolling the manufacturing supremacy of a new $30 million plant in Calgary, chock full of high-tech wizardry that was said to put the company at the head of the industry. That state-of-the-art facility sits idle today, another example of faulty planning for a market that wasn't there that seems to have plagued Novatel.[64]

Figure 18: Alberta's helping hand to high tech

General Systems Research	Magnesium Co. of Canada
The Edmonton-based laser development computer company received more than $34 million grants and other government concessions between 1984 and 1990, including a $9.4 million Alberta government loan guarantee. General Systems was put in receivership in early 1990 when that loan was called by the bank. It ultimately was sold for just $1.4 million.	The company, owned equally by the government and Alberta Natural Gas Co., received a $103 million provincial loan guarantee in 1986 to conduct experimental research aimed at turning magnesite ore into a metal stronger and lighter than aluminum. Behind on its loan payments, the company was shut down in April. The government is looking for a buyer for its High River plant.
Amptech Corp.	**Myrias Research Corp.**
The Calgary-based company makes plastic parts for the telecommunications, electronics, aerospace and defence industries. Vencap Equities Alberta Ltd., which received $200 million in government start-up funds, has invested more than $3 million in the company. Amptech profit in 1990 was $1.3 million on sales of $11 million.	Through Vencap Equities, the Alberta government held $7.2 million worth of shares in the Edmonton-based super-computer company, which went into receivership in Oct., 1990. The company ceased operation later in the year.

Source: *Globe and Mail*, May 28, 1991, p. B7.

---

64 *Ibid.*

Appendix 6A: An Exchange of Faxed Messages (Excerpts)

Fax August 6, 1991 from the first secretary, Science and Technology, Embassy of Japan, Ottawa to Palda

... A brief description written by the National Institute of Science and Technology Policy (NISTEP) ...

Tax incentives

a) Incentive for increase of R&D expenditure: 20% of the exceeding amount of R&D expenditure to the highest in the past is deductible from the corporate (or income) tax (within 10% of the corporate (or income) tax)

b) Incentive for R&D of small and medium enterprises: 6% of R&D expenditure of small and medium enterprises is deductible from the corporate (or income) tax, (within 15% of the corporate (or income) tax) (this is applied selectively with (1))

c) Incentive for R&D in fundamental technology: 7% of the cost of assets acquired for R&D in fundamental technology is deductible from the corporate (or income) tax (within 15% of the corporate (or income) tax)

d) Incentive for donation to foundations or corporations with the mission of carrying out or granting R&D : donation is calculated as losses (within a certain limitation)

e) Incentive for organizing a research union: reduction of union dues as expenses and special depreciation measure of union assets

(by NISTEP, March 1990)

Fax August 9, 1991 from Palda to NISTEP, Toyko

. . . To be precise: is there a provision in Japanese R&D/innovation tax incentives for the inclusion of market research or market launching expenses necessary for new product ventures? . . .

Fax August 20, 1991 from NISTEP to Palda

. . . We are sorry to let you know that we have no provisions in Japanese R&D/innovation tax incentives which you requested in your letter.

Please apologize to us for not being of any help to you . . .

Chapter 7

"Mais quoi! il est dans l'esprit des Français d'ordonner obstinément cet avenir qui leur échappe sans cesse."[1]

Government and Innovation Abroad

THIS CHAPTER WILL ATTEMPT TO cast light on some of the relevant features of innovation policies in Australia, France, Germany, Japan, the United Kingdom and the United States. Relevant, that is, to the Canadian case. It will briefly describe each country's tax and subsidy measures and, where information is available, procurement policies. It will also mention international cooperative efforts, and will attempt to provide short evaluations of the success or otherwise of government support of innovation. Remember, however, that the whole complex of policies toward innovation is subject to constant and rapid change and so its description is always out of date.

This chapter will be anchored by two organizing principles. First, as we have discussed already quite amply, the justification of taxpayer

---

1   George Duhamel, *Chronique des Pasquier*, Paris: Mercure de France, 1947, p. 29.

support for innovation rests on the possibility of market failure in the guises of inappropriability, large minimum efficient size, and uninsurable risk. The second organizing principle is the type of support policy: direct subsidy, tax stimulants, technological information diffusion facilities, intellectual property rights and, possibly, government purchasing policies.

A similar table (Table 35) could also be drawn up for a given industry across several economies, or even for a strategic trade objective.

In practice, however, such a grid is quite difficult to fill out without particularly detailed information. From outside of the country in question one typically has to rely on OECD publications in which data are not structured in this way. In Canada, this writer, together with a colleague, has made a start at filling out a few of the grid's windows in a recent working paper. Table 36 is the illustration taken from it.[2]

Table 35: Policies in Support of Innovation in Country X

Type of Policy	Market Failure		
	Inappropriability	MES	Risk
Direct Subsidy			
Tax stimulants			
Property rights			
Government			

---

2   Petr Hanel and Kristian Palda, "Appropriability and Public Support of R&D in Canada," School of Business, Queen's University, School of Business Working Paper 89-18, 1989.

Table 36: An Approximate Classification of Federal Expenditure on R&D in the Natural Sciences and Engineering, Fiscal 1986-1987

	Market failure		Technology for government outputs	Other
	Inappropriability	MES		
National Research Council **$415 million**	50%	30%	10%	10%
Agriculture Canada **$372**	66%	34%		
Energy, Mines, Resources **$264**	70%			30%[a]
Natural Science and Engineering Research Council **$251**	84%			16%[a]
National Defence **$220**			100%	
Regional Industrial Expansion **$186**	50%			50%[a]
Medical Research Council **$145**	90%		10%	
Fisheries and Oceans **$128**	71%			29%[a]
Atomic Energy Canada Limited **$126**	50%			50%
Environment **$63**			100%	
Communications **$63**			100%	

Total $2,234

[a]Research on behalf of industry or subsidy to industry without apparent market failure or public goods justification.

Source: Hanel and Palda, op. cit.

Table 37: Government-financed expenditure on research and development in the enterprise sector by industry group.
1985 (as a percentage of total R&D expenditure)

	Elec-trical	Chem-ical	Mach-inery	Aero-space	Other Trans-port	Chem-ical Linked	Basic Metal	Serv-ices	Total
United States	40.3	8.5[a]	13.8	76.2	14.3[e]	11.1[h]	26.4[j]	52.1	33
Japan	1.0	0.8	0.6	9.3	4.4	0.7	1.3	3.8	1
Germany	15.6	3.3	7.9	62.0[c]	2.5	9.1	19.0	43.5	15
France	32.6	5.2	14.2	62.0	3.2	2.3	3.9	11.8	23
Italy	18.9	6.7	19.6	42.2	15.0	2.9	10.9	12.4	16
United Kingdom	29.6	0.8	21.0	62.7	3.8	4.7	6.2	11.0	23
Canada	7.2	1.8	2.4	29.9	8.1[f]	5.5	2.8	18.1	10
OECD	26.8	2.5[b]	11.4	73.3[d]	5.4[g]	2.8i	5.7[k]	29.8	23

(a) 1980
(b) Excluding the United States. Approximately 5.5 percent inclusive of the U.S.
(c) 1983.
(d) Excluding Germany
(e) 1980.
(f) 1971.
(g) Excluding the United States and Canada. Approximately 9.8 percent inclusive of the United States and Canada.
(h) 1980.
(i) Excluding the United States. Approximately 7.1 inclusive of the United States.
(j) 1983.
(k) Excluding the United States. Approximately 14.0 percent inclusive of the U.S.

Source: Robert Ford and Wim Suyker, "Industrial Subsidies in the OECD Economies," *OED Economic Studies,* No. 15, autumn 1990.

The fact that such a classification cannot as yet be undertaken weakens the analytical underpinning of questions such as "Is this particular support justified from the taxpayer's viewpoint?" In general governments—taxpayers—are quite generous in their support of R&D. This is evident from Table 37 which shows both the total share of the enterprise sector financed out of public funds in the G-7 countries (last column of the table), but also a breakdown by industrial sectors. The aerospace industry stands out as the most heavily subsidized by grants and contracts.

However, not included in these figures are tax advantages offered and taken, which, in Canada, are probably three or four times the value of grants and contracts. Table 38, while unfortunately limited to the Common Market countries and encompassing all support and not just R&D, shows the relative importance of non-grant supports, such as soft loans and, of course, tax concessions.

Table 38: Support to manufacturing in EC countries by instrument[a] 1981-1986 (as a percentage of total support)

Country	Grants	Tax con-cessions	Equity partici-pation	Soft loans	Guaran-tees
Germany	35	58	0	6	1
France	20	11	26	38	5
Italy	68	11	18	3	0
United Kingdom	69	4	18	6	1
Belgium	47	2	2	10	13
Denmark	43	0	1	52	3
Greece	95[b]	0[b]	0	0	5
Ireland	39	49	8	2	1
Luxembourg	57	4	35	4	0
Netherlands	60	25	1	13	0
EC-10	47	23	14	14	2

[a]EC countries: excluding supranational support.
[b]Tax expenditures included in grants.

Source: Same as Table 37.

Tax Stimulants

We start with a comparison of tax stimulants which draws in part upon the work originated a decade ago by McFetridge and Warda and updated recently, under the auspices of the Conference Board, by Warda.[3]

Table 2, taken from the latter's study, showed the so-called B-index, a measure of a country's (plus province's or state's) generosity toward R&D entrepreneurs in 10 jurisdictions. Table 39, also taken from Warda's publication, shows the tax allowances in a more detailed and "structural" manner.

In the **United States** the tax picture is quite straightforward.[4] As in all the other listed countries *current* R&D expenditures are 100 percent deductible in the year expended, with the possibility of amortization over 60 months. (In Canada deductions can be deferred to the future). *Capital* expenditures on equipment, however, are not immediately deductible as in Canada and will be generally depreciated over 5 years. The U.S. federal government offers a 20 percent investment tax credit on the increase in current R&D expenses over the average of R&D expenses for three prior years. This investment credit reduces the deductible base for current R&D expenditures by 50 percent of its value. (In Canada, it is taxable the following year). A congressional battle is fought every year about the provision of the tax credit, which has been on the books since about 1984. So far R&D has always won. This periodic review of the tax incentive apparently makes long-range R&D planning in the United States more difficult than in Canada.

The Deloitte & Touche report prepared for ISTC in May, 1990 contains two tables, which have been merged and are shown here as Table 40. This book's table 31, also from that report, shows Canadian R&D tax incentives. Comparing table 31 with table 40, the Deloitte & Touche paper states that the Canadian tax system clearly offers more

---

3   Donald G. McFetridge and Jacek P. Warda, *Canadian R&D Tax Incentives: Their Adequacy and Impact,* Toronto: Canadian Tax Foundation, 1983; Jacek Warda, *International Competitiveness of Canadian R&D Tax Incentives: An Update,* Ottawa: Conference Board of Canada, May 1990.

4   The description of Canada's tax laws in respect of R&D is in Chapter 6.

Table 39: Tax Treatment of R&D—Canada vs. Major Industrial Countries

| | Corporate Income Tax (%) | Present Value | | | Tax credits (allowances) | | |
| | | Current R&D deduction per dollar | Capital cost allowance per dollar | | Percent | Increment (I) or level (L) | Current capital or both |
			Machinery	Buildings			
Canada-Federal	39.8[a]	1	1	<1	20	L	Both except buildings
Canada-Ontario	40.3[a]	1	1	<1	25/37.5	Both	Both except buildings
Canada-Quebec	32.0[a]	1	1	<1	20	L	R&D Wages
Canada-Nova Scotia	40.8[a]	1	1	<1	10	L	Both except buildings
U.S.-California	40.1	1	<1	<1	20/8	I	Current
U.S.-Illinois	36.6	1	<1	<1	20	I	Current
Australia	39.0	>1	>1	<1	n.a.	n.a.	n.a.
Japan	50.75	1	<1	<1	20	I	Both except buildings
Korea	39.75	1	<1	<1	10	L	Both
France	39.0	1	<1	<1	50[b]	I	Both except buildings

Table 39: (continued)

	Corporate Income Tax (%)	Present Value			Tax credits (allowances)		
		Current R&D deduction per dollar	Capital cost allowance per dollar		Percent	Increment (I) or level (L)	Current capital or both
			Machinery	Buildings			
F.R.Germany	56.0	1	<1	<1	7.5	L	Capital
Italy	46.37	1	<1	<1	n.a.	n.a.	n.a.
Sweden	52.0	1	<1	<1	n.a.	n.a.	n.a.
United Kingdom	35.0	1	1	1	n.a.	n.a.	n.a.

n.a.—Not available.
[a]Current tax rates; from July 1, 1991, these rates will be reduced by 2 percentage points due to a full phase-in of the manufacturing and processing tax deduction (5 per cent).
[b]In addition, firms that did not perform R&D prior to 1987 can claim a volume tax credit of 30 percent on the excess over 1987 spending levels.

Source: Jacek P. Warda, *International Competitiveness of Canadian R&D Tax Incentives: An Update*, Ottawa: Conference Board of Canada, May 1990.

Table 40: After-tax Costs for R&D Expenditures in the USA, 1990

	For Small R&D Performers	For Large R&D Performers
R&D Expenditure	$1,000	$1,000
Federal Investment Tax Credit (20% x 0.5 ($1,000))	(100)[a]	(100)[a]
Tax Saving from deduction (37%[b] of (1,000 - $100)) (40%[c] of (1,000 - $100))	(333)	(360)
After-Tax Cost	$567	$540

[a] Qualifying base period expenses must be at least 50% of the current year qualifying expenditures. Only $500 of the $1,000 in R&D expenditure is qualifying R&D expenditures for tax credit purposes.

[b] 37% is an estimated combined federal and state income tax rate for U.S. companies with taxable income of $200,000 U.S. per annum.

[c] 40% is an estimated combined federal and state income tax rate for U.S. companies with taxable income above $335,000 U.S. per annum.

Source: Deloitte & Touche, *A Comparison of Tax Incentives for Performing Research and Development in Canada and in United States*, Ottawa: May, 1990.

generous provisions to R&D performers. It goes on to say that there are additional factors that make the Canadian system more generous:

1. The ability to write off capital expenditures (on equipment, not on buildings) immediately rather than over their useful life as under U.S. rules.

2. Greater flexibility in determining when one writes off R&D expenditures.

3. In certain cases, the refundability of the R&D tax credit in Canada versus non-refundability in the U.S.

4. The R&D tax credit in the U.S. is for incremental expenditures only and therefore is of far more limited value than the Canadian R&D tax credit.

A host of other minor items is also more advantageous in Canada, such as that only 65 percent of extramural R&D is eligible for the tax credit in the USA. Deloitte & Touche, as well as other sources, do point out, however, that the audit process in Canada is much more strict and time consuming: in every instance a tax return must be audited and it must be established that "scientific research and experimental develop ment" underlay the submitted expenses. Other sources, authoritative but "not for attribution," point out that only first-time applicants are subject to an audit without fail.

In **France**, as is evident from Table 39, capital R&D expenditures are not fully deductible in the year incurred, though are depreciable within three years. The tax credit is not taxable—a definite advantage—and is equal to 50 percent of the increase in R&D expenses over the preceding year. In the budget year 1991 the ceiling of the tax credit applicable to a firm was raised from 10 to 40 million francs and the eligible fixed R&D expenses were raised to 65 percent.[5] It was expected that by 1992 tax credits taken up will amount to F3.8 billion.

In **Germany** only current R&D expenditures are deductible, with capital expenditures given the same treatment as other depreciable assets. But a tax credit is granted on fixed assets used for R&D purposes with a rate of 7.5 percent for investments in excess of DM 500,000. This tax credit is estimated to have yielded DM 450 million in tax reductions to firms in 1989.[6] A special R&D depreciation provision yielded about DM 255 million in tax abatements to enterprises in the same year. A three-stage tax reform 1986-1988-1990 is apparently eliminating special-

---

5 Ministère de la rechereche et de la technologie, *Le budget civil de R et D pour 1991*, Paris: 1990.

6 Bundesministerium fur Forschung und Technologie, *Faktenbericht 1990 zum Bundesbericht Forschung*, Bonn: March 1990, Table I/14.

ized tax exemptions and instead lowering the overall federal taxation rates.[7.]

In **Australia** the principal means of taxpayer support to innovation since 1985 has been a 150 percent deduction for current R&D expenditures. Equipment investment depreciation (3 years straight-line) also benefits from the 150 percent provision. This tax feature is to continue till mid-1993, with 125 percent for another two years.[8] In the words of the OECD, Australian "authorities report that the tax incentive has helped to stimulate an 81 percent real increase in business sector performance of R&D, from 0.34 percent of GDP in 1984-5 to 0.52 percent of GDP in 1988-9."[9] We should not take their word for granted, as will be discussed later on.

The **United Kingdom**'s taxation policies do not single out R&D for special treatment. As can be observed, however, in Table 39, the UK has the lowest corporate tax rate of the ten countries compared.

Finally, in **Japan**, as already mentioned in Appendix 6A, tax incentives—as far as can be gleaned from the direct communication with Japanese officials—are not particularly different from those offered in the other countries covered here. There is a reward for increased R&D expenditure, a stimulant to small-business research and to "charitable" donations to R&D, and an incentive to carry out fundamental technological research. No tax credit is available.

The Effectiveness of Tax Stimulants

Independent scholarly evaluations of tax stimulants tend to center on tax credits and special allowances rather than tax deductions or capital equipment write-offs. This is presumably so because, in general, R&D can be as fully expensed as any other business outlay and its capital investment features are given similarly favourable treatment. Attention

---

7   *Ibid.*

8   OECD, *Industrial Policy in OECD Countries, Annual Review 1990*, Paris: 1990, p. 24.

9   OECD, *Industrial Policy in OECD Countries, Annual Review 1991*, Paris: 1991, p. 28.

is therefore paid to the "extra" enticement which is usually offered in the form of a tax credit or special allowance or concession.[10]

As Dwyer points out in a particularly lucid piece devoted to the appraisal of the Australian 150 percent tax concession, an informed assessment of the impact of a tax stimulant requires several items of information:

- the influence on the user cost of R&D
- the responsiveness of R&D expenditure (elasticity) to changes in user cost
- the revenue foregone by the taxpayers
- the social benefits from R&D
- the allocation of firm funds between R&D and investment[11]

The comparison of R&D *user costs* is carried in the B-index table 2. Canada ranks first in cost reduction. The *responsiveness* of R&D expenditures to tax stimulants in the *United States* and *Canada* has been appraised by Mansfield and Switzer and by Bernstein, as mentioned in chapter 6 under the heading "how effective has government support been?" The latter estimates that a dollar of tax incentive increases R&D outlays by 80 cents. The former believe that the response elasticity is only 0.3 or 0.4, or about 30 or 40 cents on the dollar.

This writer does not know of any published studies regarding similar elasticities in *Japan, France* or *Germany*. One can merely infer that they may be unsatisfactory in Germany, given that the tax reform carried out has as its purpose a general lowering of overall tax rates, supported by the elimination of special concessions.

In *Australia* Dwyer believes that in the three fiscal years 1985-6 to 1987-8 during which the 150 percent concession was on the books, gross R&D expenditure increased by 47 percent, considerably less than the 81

---

10 Special allowances or tax concessions, as they are sometimes called, allow extra *deductions* for R&D, beyond the deduction rate for ordinary business expenses. The best example is that of the 150 percent tax concession in Australia. Tax credits are directly applied against income tax payable. Both kinds of tax stimulants obviously reduce the corporate income tax burden.

11 Larry Dwyer, "The Impact of the 150 Percent Tax Concession for Industrial R&D in Australia—A Preliminary Assessment," *Prometheus*, December 1989, pp. 316-332.

percent increase ascribed to this measure by the earlier cited "authorities." Just as in other countries that favoured R&D by tax abatements, there was a distinct possibility that other business expenditures were redefined to fit the R&D definition. Adding to a possible overestimate of the response is the fact that not every R&D-spender can benefit from the tax concessions given that there may be no tax liability in the first place.

The *tax revenue foregone* in favour of R&D/innovation concessions should, in general, be estimated in annual government budgets. But in the case of tax credits or, as a matter of fact, in the more general case of any allowable R&D deductions, budgets do not often reveal the cost to the taxpayer. These measures do not entail any disbursements—except in the Canadian case—and the tax costs are more difficult to estimate than budget outlays, since not all firms will take advantage of the concessions.

Estimates of foregone revenue due to tax credits for Canada are given in Table 28. In Australia, according to Dwyer, the ratio of incentive-induced R&D spending to the tax revenue given up by the federal government is about 0.7 or less, as indicated in Table 41.

Table 41: Maximum Estimates of a Ratio of Incentive-Stimulated R&D Spending to Government Revenue Foregone

	1985-1986	1986-1987	1987-1988
Est. induced R&D outlays ($M Australian)	52	120	134
Revenue foregone ($M)	87	170	200
Ratio	0.6	0.7	0.7

Source: Dwyer, *op.cit.*, table 4.

The *social benefits* of tax concessions as compared with the tax costs incurred (as in the cost-benefit formula specified in Chapter 6) have not been widely investigated. The idea that they exist is based on some research which found evidence of spillovers and of private returns to R&D higher than those to other capital investments.[12] If, therefore, R&D tax concessions do stimulate research, and research has both high internal and external returns, stimulative R&D policies paid for by the taxpayer must have excellent returns, even if one dollar of tax concession is far from generating one dollar of additional research expense. This bold assertion we need not take for reality, particularly if we remember some of the opportunity costs of raising government revenue which were outlined in the preceding chapter.

If the B-index is regarded as the minimum benefit-cost ratio at which an R&D project is profitable for the firm, then a similar index based on the tax treatment of non-R&D capital investments can be used to evaluate the possible *re-allocation of the firm's funds from investment to research*. Since B-indexes for R&D are usually higher than those for capital investment, that is, since tax concessions to research are more generous than those to non-R&D investments, it is likely that the introduction of additional stimulants to research will shift funds from capital investments. Dwyer estimates that the Australian B-index stood, after its introduction in 1985-6 at 0.62 while the capital investment index was 1.25.[13] The probable outcome was that firms with R&D capabilities undertook research projects with benefit-cost ratios between 0.62 and 1.25 while rejecting capital outlays with benefit-cost ratios within this range. For that year Dwyer estimates that capital expenditure could have been reduced as much as by $106 million (Australian), for the benefit of R&D outlays. We do not have similar information for other countries though the data to perform similar calculations are obtainable.

---

12   Jeffrey I. Bernstein's articles cited in Chapter 2.

13   Dwyer, *op.cit.*, p. 327.

Subsidies to Industrial Innovation

Taxpayers, through their elected representatives and through appointed bureaucracy, support both basic research in universities and various institutes and government laboratories as well as (mostly) applied research and development in the for-profit sector. This section tends to focus on industrial subsidies, given our preoccupation with industrial innovation policies. We cannot, however, exclude totally glimpses of what is commonly called "science policy." While R&D/innovation subsidies are often included in the more general subsidization of industry, we can only rely on description or statistics dealing with

Table 42: Company and Federal Funding of Industrial R&D for Selected Industries, 1981 Estimates (in millions of dollars)

Industry	Total	Federal	Company
Total[a]	57,719	17,362	40,359
Chemicals and allied products	5,325	383	4,942
Petroleum refining and extraction	1,920	140	1,777
Rubber products	800	190	616
Primary metals	889	182	707
Fabricated metal products	638	80	558
Nonelectrical machinery	6,800	739	6,061
Electrical equipment	10,466	3,962	6,502
Motor vehicles	5,089	704	4,381
Aircraft and missiles	11,702	8,501	3,201
Professional and scientific instruments	3,685	638	3,047
All other manufacturing	8,325	963	7,368
Nonmanufacturing industries	2,080	880	1,199

[a]These totals differ from that of the source.

Source: From Table 6.9, p. 141, in David D. Mowery and Nathan Rosenberg, *Technology and the Pursuit of Economic Growth*, Cambridge University Press, 1989.

explicit support to the development/diffusion of new products or processes.

Surprisingly, in the *United States* there are hardly any subsidies to private-sector industrial research listed.[14] How is it, therefore, that one-third of all industrial research is government-financed in the USA, as seen in Table 37? The bulk of the answer lies in the system of research contracts entered into by the government with private corporations. Most of the contracts are for defence-related research, as can be seen in Table 42 which shows that the federal contribution to aircraft and missiles represents, by itself, almost 50 percent of all federal subsidies.

Presumably a procurement contract is for the hardware itself and for a cost-plus research undertaking by the government's supplier.[15] One can therefore conclude that the government's *procurement* policies are an important element in stimulating technological advance among private-sector suppliers.

Mowery and Rosenberg discuss this issue extensively, adducing both pros and cons for this view.[16] Among their "contra" arguments is the interesting one which speculates that product designers working on defence projects may actually learn both to "overdesign" to excessively high qualities and to learn an indifference to cost, which may serve their companies poorly when they attempt to apply the new technology to civilian markets. On the other hand there is the importance of assured demand during start-up periods in civil aviation, electronics, and computers—a demand springing from defence needs—a view also shared by Nelson.[17]

---

14 One loud, but dollar-small exception is Sematech, described later on.

15 McFetridge and Warda, *op.cit.*, p. 80, argue persuasively that the element of hidden subsidy in such contracts is negligible.

16 David C. Mowery and Nathan Rosenberg, *Technology and the Pursuit of Economic Growth*, Cambridge University Press, 1989, pp. 140-150.

17 Richard R. Nelson, "Government Stimulus of Technological Progress: Lessons from American History," in R.R. Nelson (ed.), *Government and Technical Progress: A Cross-Industry Analysis*, New York: Pergamon Press, 1982, p. 481ff.

Promotion of basic and pre-competitive non-defence research in the United States is largely the domain of the National Science Foundation (NSF) and the Technology Administration (TA). NSF supports basic research through Engineering Research Centres (19 operating, first established in 1985), Industry-University Cooperative Centres (41 now operate, first established in the late 1970s with participation of over 300 firms), Advanced Scientific Computing Centres (four currently operating) and Interdisciplinary Science and Technology Centres (eleven were established from 1989 on operating in areas such as superconductivity, robotics, biotechnology and microelectronics) as well as nine Material Research Laboratories. The Technology Administration was created in 1988 in the Department of Commerce to co-ordinate and undertake science and technology initiatives, combined with the promotional activities of the National Institute of Standards and Technology (it has a direct mission to assist and support US industry, for example, with new outreach programmes to assist R&D in smaller firms), and the national Technical Information Service. The Technology Administration promotes and removes barriers to collaborative R&D, as well as re-orients the Advanced Manufacturing Program.[18]

One part of the activities described above can be viewed as implementation of government science policy. Another part, overlapping the first, is clearly designed to foster the *diffusion and adoption* of innovations.

The mention of pre-competitive research evokes the government-subsidized scheme called Sematech. Box 10 gives a view of this curiously un-American venture. The European JESSI venture in semiconductors, described at the end of this chapter, bears similarities to Sematech.

The OECD estimates that in 1990 *Australia's* support for industrial R&D relied on tax concessions to the extent of 70 percent, while grants or subsidies constituted about 30 percent of the treasury's outlay. Table 43 gives a glimpse of the variety and of the amounts of direct subsidies channelled to private business.

---

18   OECD, *Industrial Policy in OECD Countries, Annual Review 1990*, Paris: 1990, p. 24.

Box 10: SEMATECH—Semiconductor Manufacturing Technology

This narrative is based on Brink Lindsey's article "Dream Scam" in the magazine *Reason*, February 1992, pages 40-45.

At the beginning of 1987 the U.S. Defense Science Board reported that "it is simply no longer possible for individual U.S. Semiconductor firms to compete independently against world-class combinations of foreign industrial, governmental and academic institutions." Thus "a direct threat to the technological superiority deemed essential to U.S. defense systems exists." In March 1987 14 U.S. chipmakers formed a consortium to face the Japanese challenge in developing advanced manufacturing techniques.

Since Sematech represents firms in a high technology industry, and not in a declining smokestack sector, and since it is dedicated to "precompetitive" R&D, and not to picking winners and losers, this consortium's public support should be an ideal industrial policymaker's target.

Sematech's mission to develop low-cost effective manufacturing techniques in a specially constructed large chip factory in Austin, Texas to all purposes failed. Its "precompetitiveness" character precluded selling those *memory* chips (DRAM's-dynamic random access memory) and so achieving the long production runs that permit dramatic learning curve cost decreases. Sematech then turned to helping American chipmakers indirectly, by assisting American chip manufacturing equipment suppliers who were also besieged by the Japanese.

In the meantime, however, the market has increasingly turned to logic chips which manipulate data rather than just store them-the microprocessor is the best example here, which acts as the central processing unit in personal computers. The logic chips are custom-manufactured and the premium here is on design rather than on the manufacturing process. American companies, especially nimble newcomers with their creative design strength, hold now close to 50 percent of this rapidly growing market which is presently one-third larger than the declining one of commodity memory chips.

In Lindsey's opinion

> Buzzwords like *precompetitive* notwithstanding, Sematech is yet another example of government meddling in an industry to pick winners and losers. And as usual, the bureaucrats have backed the wrong horse. Within the sunrise industry of micro-electronics, the government has managed to locate and subsidize the sunset companies, to the detriment of those young and dynamic companiees that represent the industry's future.

*With permission from the February 1992 issue of **Reason** magazine. Copyright 1992 by the Reason Foundation, 400 – 3415 S. Sepulveda Blvd, Los Angeles, CA 90034.*

Table 43: Australian Industrial Research and Development Incentive Grants, 1978-79 to 1984-85 (Thousands of dollars)

Year	Commencement grants	Project grants	Public interest projects	Total
1978-79	6,000	16,500	1,249	23,749
1979-80	7,000	22,950	3,999	33,949
1980-81	9,657	36,056	5,000	50,714
1981-82	9,700	12,053	2,445	24,198
1982-83	13,075	34,797	4,935	52,807
1983-84	14,558	43,243	8,138	65,939
1984-85	16,262	38,126	9,822	64,210

Source: From Table 1 in Stuart Macdonald, "Theoretically Sound: Practically Useless? Government Grants for Industrial R&D in Australia," *Research Policy*, 15 (1986), pp. 269-283.

Project grants provide the main component of the grant program. The granting board considers the relative technical and commercial merits of the submitted project; it is apparently neutral between technologies and industries, but did, at one time or another favour pre-competitive R&D in specific generic technologies.[19] As with most grants, they are limited as to the percentage of the project they finance and as to the total amount (50 percent and $750,000 in 1985).

Commencement grants to encourage firms to engage in R&D have since been replaced by the grants-for-industry (GIRD) scheme to pro-

---

19   Australian Industrial R&D Incentives Board, *Future Government Support for Innovation*, Canberra 1985 and OECD, *Industrial Policy in OECD Countries, Annual Reviews 1990* and *1991, op. cit.*

vide an incentive to smaller younger firms which may be innovative, but cannot benefit from the tax concessions since they are not yet profitable.

Public interest grants were providing substantial support for specific projects of exceptional merit considered in the public interest. These descriptions provide a mid 1980s snapshot of Australian support schemes. They appear to be as complex, bureaucratically infested, and changeable as the Canadian ones. It is therefore not surprising that the annual OECD reports, which started in 1986 and deal with industrial policies, always carry the lengthiest descriptions with respect to Australia.

Therefore, it is also no accident that in Australia substantial attention is devoted to appraising government innovation forays. A typical example is an article by Stuart Macdonald which is a sombre critique of Australian support schemes.[20] Macdonald's article carries a table, based on a survey of 14 Australian firms who received $8.6 million in grants between 1978 and 1980, representing 22 percent of all project grants allocated during this period. While such grants were given on condition that they be matched by company funds, 8 firms subtracted $2.6 million from their normal R&D budget, 2 firms added the grants to the budget, 3 firms matched the grants to the extent of $5 million and 1 more than matched a $400,000 subsidy. In this case, observes Macdonald, government grants had no more than one-third of their assumed incentive effect. It seems quite common to regard an R&D grant as windfall income. The Australian experience appears therefore quite close to Canada's, as described in chapter 6, referring in particular to footnotes 10 and 28 therein.

We should also mention that a National Preference Agreement exists which is aimed at coordinating state purchases with the federal government's *purchasing policies* to stimulate new technology. Finally, Australia is launched on a massive program of technology diffusion. Established in 1990, the Cooperative Research Centres Programme grew

---

20 Stuart Macdonald, "Theoretically Sound: Practically Useless: Government Grants for Industrial R&D in Australia," *Research Policy*, 15 (1986), pp. 269-83.

out of a concern over inadequate linkage between researchers in different institutions and between research and its applications. 50 Centres are to be set up, with financing of A$100 million a year.[21]

In *Germany* in recent years emphasis has been put on public support to basic research, on the technological upgrading of small and medium enterprises (SMEs), and on international cooperation. As can be seen in Table 44 the grant portion of support to industrial R&D amounts to 90 percent of the taxpayers' effort. Figure 19 shows federal outlays (Laender financing is almost as important) for the Max Planck Gesellschaft

Table 44: Government policy instruments used to support industrial R&D: 1985-1986 (approximate share of 1985-1986 expenditures in brackets)

Canada	Grants (100%) (tax concessions not included)
Japan[1]	"Consignment" subsidies (40%), tax concessions (35%), grants (25%), equity capital (from 1985) (2.5%)
France	Grants (50%), repayable grants (25%), tax concessions (from 1984) (25%)
Germany	Grants (90%), tax concessions (10%)
United Kingdom	Grants (65%), mixed grants + loans (35%)
Australia	Tax concessions (from 1985/86 70%), grants (30%)

[1]"Consignment" subsidies involve research commissioned by central government from private industry associations, groups of private firms and government laboratories in national co-operative research projects such as the Large-Scale Projects, or the Fifth-Generation Computer Project.

Source: OECD, *Industrial Support Policies in OECD Countries 1982-1986*, Paris: 1990.

---

21   OECD *Industrial Policy in OECD Countries, Annual Review 1991, op. cit.*

Figure 19: 1986 to 1990 Actual and Estimated (Soll) Outlays of the German Federal Government in Support of MPG, DFG and FhG Programs (Millions of DM)

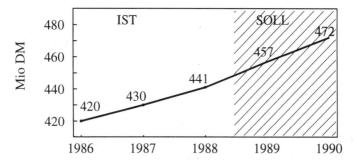

FuE-Ausgaben des Bundes fur Grundfinanzierung MPG

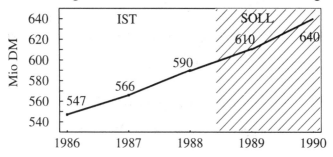

FuE-Ausgaben des Bundes fur Grundfinanzierung DFG

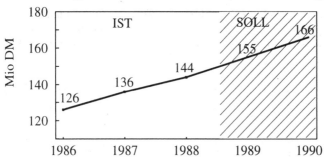

FuE-Ausgaben des Bundes fur Grundfinanzierung FhG

Source: Bundesministerium fur Forschung und Technologie, *Schwerpunkte der F-u.-E. Forderung des Bundes*, Bonn: 1990, pp. 91-92.

Figure 20: Research and Development Expenditures in the German Federal Republic by Funding Sectors, 1981-87 Actual, 1988 and 1989 Estimated (Billions of DM)

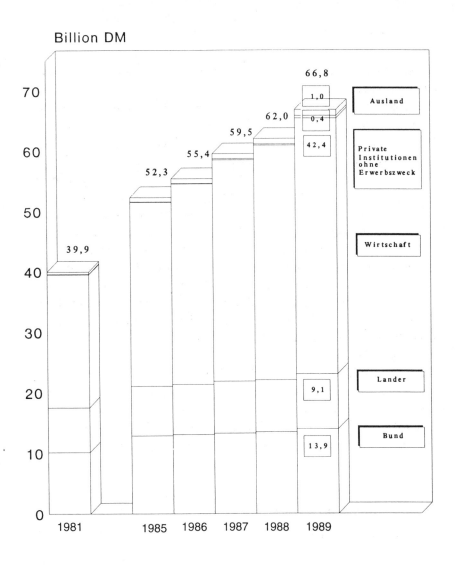

Translation: Bund = federal state; Länder = provinces; Wirtschaft = private sector; PIoE = private non-profit; Ausland = from abroad.

(MPG) which finances basic research institutes; for the Deutsche Forschungsgemeinschaft (DFG) which funds university research; and for the Fraunhofer Gesellschaft (FhG) that finances a string of applied research institutes.[22] Figure 20 gives another detailed look at the financing of German R&D by the federation (Bund), the Laender, and the private sector.

From 1990 onwards a deliberate effort has been undertaken to bring research and technology up to snuff in the eastern Laender. This is done, among other things, by providing grants for research personnel and R&D staff, by stimulating research in SMEs, and by promoting technology transfer and diffusion via technology parks, business incubators and joint industrial research activities, as well as by establishing information centres in new technological fields.[23]

The *United Kingdom*, as documented in Tables 39 and 44, appears to rely on subsidies to support innovation to an even greater extent than Germany. The emphasis on support to defence/nuclear-oriented industries substantially diminished in the 1980s and a greater emphasis was placed on the promotion of pre-competitive cooperative projects, both domestic and international, and on assistance to small and medium enterprises.[24] The encouragement of collaboration between companies and scientific institutions, including universities, is the task of the Link program. As Box 11 illustrates, it is oriented both to innovation creation and its diffusion or technology transfer.

As of the end of 1990 there were 27 programs under this scheme, with a total value of £350 million, including both taxpayer and private funding. The ventures were mainly in biotechnology, advanced materials and advanced manufacturing technology.[25] Box 12 describes the SMART program that promotes innovative projects in the SMEs. Critics of the program might point out that if the Department of Trade and

---

22   OECD, *Structural Adjustment and Economic Performance*, Paris: 1987, p. 106.

23   OECD, *Industrial Policy in OECD Countries, Annual Review 1991*, p. 28.

24   OECD, *Structural Adjustment, op. cit*, p. 106.

25   See *OECD, ibid, p. 31*.

Box 11: United Kingdom LINK—Collaborative Research "Bridging the Gap Between Science and the Market Place"

LINK is an initiative for the support of strategic, i.e. long-term, or pre-competitive industrially relevant research, involving companies and science base institutions in collaborative research projects. It will involve all of the major R&D Departments of Government and Research Councils through support for programmes in their own spheres of interest.

LINK aims to accelerate the commercial exploitation of government-funded research. The initiative will focus on advances in science and engineering with particular commercial promise. It will stimulate collaboration between industrial and science base partners on projects in key areas of science and technology. This will speed the future exploitation of new processes, and the development and profitable marketing of new products and services by industry.

The objectives of LINK are to:
- foster priority areas of research;
- stimulate an increase in industry's investment in R&D;
- help industry exploit scientific developments;
- make scientists and engineers more aware of industry's needs;
- develop technologies which cross industrial sectors/scientific disciplines.

The benefits of LINK are:
- researchers will see their work successfully developed and will become better informed of industry's needs;
- industrialists will benefit from the wealth of expertise in establishments of higher education and research, thus improving their competitive position both at home and in the world's markets;
- the economy will benefit because industry will be better able to grasp new opportunities at the cutting edge of international competition.

Link provides a framework for collaborative, pre-competitive research involving the science base and industry. It aims to help industry exploit developments in science, by fostering priority areas of scientific research services. It also aims to stimulate industrial investment in R&D (the government is providing £210 million over the first five years, to be matched by at least as much from industry).

LINK resources are being concentrated on a series of programmes in areas of science and technoloy with particular promise in terms of the objectives of LINK. Each program will consist of a portfolio of collaborative projects.

The LINK initiative reflects the trend towards government support for collaborative, pre-competitive research, and away from support for single firm near-market product development. Universities and other Higher Education institutes will be major players in LINK as the science base partners in projects. LINK is a key element in the government's strategy for the support of long-term research which has potential for exploitation by industry, but which needs further development before commercial applications.

Source: OECD, *Industrial Policy in OECD Countries, Annual Review 1989*, Paris: 1989, p. 26.

Box 12: Promotion of R&D in Small and Medium-Sized Enterprises—The "Smart" program in the United Kingdom

SMART (Small firms Merit Award for Research and Technology) is a competition in two stages which provides support for innovative technological ideas. Funded by the Department of Trade and Industry (DTI), it is open to individuals and businesses with fewer than 50 employees. Through SMART, the DTI intends to demonstrate the commercial potential of innovative firms and show private sector institutions that their support as providers of finance is worthwhile.

The aims of the award are:
- To bring forward highly innovative projects which have commercial potential but are now dormant because existing sources of finance do not wish to support them;
- To encourage the formation of small firms which will develop and market new ideas in selected areas of science and technology;
- To help these small firms to mature sufficiently for private sources of funds to take a practical interest.

Following a pilot exercise in 1986, the first full competition was held in 1988 when 140 Stage 1 awards were made. In 1989 a three-year series of annual SMART competitions was launched with £29 million funding. The number of Stage 1 awards has risen to 150 in 1989 and 180 in 1990. It is expected that about half of Stage 1 winners will move on to Stage 2 a year into their projects. Stage 1 is concerned with studying the feasibility of an innovative proposal and Stage 2 with further development e.g. of a pre-production prototype.

Stage 1 winners receive 75 percent of their project costs up to a maximum of £37,500; Stage 2 winners receive 50 percent of project costs up to a mazimum grant of £50,000.

The awards are made at the discretion of DTI and the selection process for Stage 1 takes account of:

- The quality and novelty of the proposal;
- The qualifications and experience, in both R&D and business, of the project leader and team;
- The significance of the project and its potential commercial benefit to the UK;
- The means proposed for turning the idea into a commercially successful product or process;
- Any contribution the project will make to urban regeneration.

During selection for Stage 2 DTI looks also at the technical progress made during Stage 1 compared with the original proposal, and at the size and structure of the market for the new product or process.

Source: OECD, *Industrial Policy in OECD Countries, Annual Review 1991*, p. 35.

Industry can determine which of the projects warrant funding, private markets may well have funded these ventures anyway.

The unsuspecting but impressed imaginary traveller who takes the Concorde from New York to Paris and the TGV from the Gare de Lyon past the Rhône nuclear power station complex to Marseille has unwittingly been exposed to three of *France's* monumental innovation subsidies. The Concorde, of which there are five or six still flying, cost the French taxpayer anywhere upward of $5 billion (in 1970 purchasing power) for development.[26] As of 1982, France provided F100 million from its *research* budget for subsidies to Air France so that it could continue to fly the planes. To this day it is likely that each Paris-New York flight does not even cover variable costs, probably to the extent of $10,000 per flight.

All the world admires the TGV. However, this writer was unable to locate a single benefit-cost analysis of its development, or of its infrastructure cost—probably comparable to those of an autoroute—nor of its environmental impact as the frequently pylon-elevated tracks crisscross *la doulce* France.

Close to 80 percent of France's electricity is supplied by its nuclear generating capacity, the highest percentage in the world, well ahead of Belgium and Ontario. EDF, Electricité de France, the French electricity monopoly, has a debt load of about F230 billion and its accumulated losses between 1974 and 1989 reached 28 billion francs.[27] Annually, the research budget of CEA, the Commissariat à l'énergie atomique, hovers around F3 billion. Framatome, the sole, state-owned supplier of nuclear power stations to EDF has been hovering, too, but on the brink of bankruptcy.

---

26 If we assume that France shared costs equally with Britain in the Concorde's development, which is a conservative assumption. See P. Henderson, "Two British Errors: Their Probable Size and Some Possible Lessons," *Oxford Economic Papers*, 1977, pp. 155-205.

27 "France: Problems of Nuclear Energy," a translation of an *Economist* article which appeared in *Hospodarske noviny*, No. 15, 1991 and *Time*, July 15, 1991, p. 31.

In the words of the OECD, in France the perception of national independence as extending to economic matters, and the special features of the training system that produces its top-level technocrats, largely explain the French preference for a technological development strategy centred on large-scale programs conceived, directed and implemented either by teams of top civil servants appointed specifically for that purpose or by large, often public enterprises working hand-in-glove with government, partly because of the close bonds uniting those who hold the reins of technology and governments.[28]

What is the size of all these subsidies? The 1990 *civilian* budget for R&D in France is set out in Figure 21. Almost 5 billion francs are allocated as direct subsidies to industrial research, but the other pie slices also contain subsidies. ANVAR (Agence nationale pour la valorisation de la recherche) is the most visible government agency through which subsidies are channelled, via regional offices, to industry for support not only of R&D, but of innovative activities further "down the line," such as marketing research.

The general mix of assistance is interesting, for close to a quarter of it comes in repayable though necessarily "soft" loans (see also Table 44). The "grands programmes" provide support for research—but not only in the private sector—for nuclear power (see the already mentioned CEA with about F3.1 billion in 1991), aeronautics (Airbus and the CFM jet engine with F2.9 billion in that year), space, Mitterand's high-definition television obsession, etc. The other slices in Figure 21 represent, among other things, a very important aid to electronics. In 1991 governmental outlays constituted about one-half of GERD (gross expenditure on R&D) which amounted to almost 2.5 percent of GDP; defence-oriented research outlays were about 20 percent. In 1989, of the F86 billion spent on research by industry, public financing amounted to F17.36 billion or 20 percent.[29]

There is considerable overlap between straightforward subsidization to industry, regional development and innovation aid. There is thus

---

28   OECD, *Structural Adjustment, op. cit.*, p. 217.

29   Various PR handouts for 1991 by the Ministère de la recherche et de la technologie.

Figure 21: France—Budget civil recherche et développement, 1990

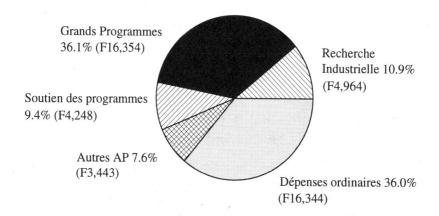

BCRD pour 1990
Dotations en millions de F.

Grands Programmes 36.1% (F16,354)

Recherche Industrielle 10.9% (F4,964)

Soutien des programmes 9.4% (F4,248)

Autres AP 7.6% (F3,443)

Dépenses ordinaires 36.0% (F16,344)

Source: Ministère de la recherche et de la technologie, *1989-1990 Un nouvel élan pour la recherche*, Paris, 1990.

a veritable non-Cartesian jumble of assistance programs which it is almost impossible to unravel, even with the aid of the OECD's authoritative (1986) report, *Innovation Policy: France*. A taste of this complexity is apparent from the admirably concise summary from the 1990 OECD annual report on industrial policy.[30]

> In France government support is provided if it has a multiplier effect on private R&D effort or increases the quality of research (for example increasing co-operation between enterprises). The nature (grant or *repayable advance*) and importance of assistance also depends on the level of risk for the enterprise and the stage of research (basic, applied, pre-competitive). Support is i) for targeted research (allowing mobilisation in priority areas); or ii) through general measures available to all firms if they meet

---

30   OECD, *Industrial Policy in OECD Countries, Annual Review 1990*, p. 115.

support requirements. General assistance is also available through the tax credit (used to encourage contract research and SME research). Policies focus on SMEs (consultancy and advice, assistance and grants, training, technology transfer). There is no sectoral aid but there are programmes to mobilise industry around strategic technologies, to *integrate new technologies in products and processes*, improve the R&D environment (tax credit, more accessible technical information, technology transfer, research training), and to increase the orientation towards European and other international R&D programmes.

One can speculate that such complexity plays into the hands of an independent decision-making mandarinate.

French Innovation and Industrial Policy—A Brief Critique and Aside

This seems as good a place as any to reflect, in this comparative chapter, on French policies toward innovation because probably more strongly than elsewhere, they are inextricably linked with overall industrial policy. From a critical point of view the best representative example of these policies is the French state's or the French bureaucracy's fostering of the electronics and computer sectors, currently known also as "la filière électronique," but by other failed epithets previously.[31]

Nowhere, as de Vos mentions in his Science Council monograph, has the full range of French planning proclivities been better illustrated than in the field of electronics.[32] In 1983 "le grand public" was made aware of the second battle of Poitiers in which the flood of Japanese imported video recorders was made to pass through the narrowest of non-tariff barrier bottlenecks in the persons of two overworked customs officials who had to check every single machine trucked from ports of entry to their small warehouse in that city. (In the first battle there in the

---

31 For more detailed descriptions of other failed attempts, see Palda, *Industrial Innovation*, Vancouver: The Fraser Institute, 1984, Chapter 6, and OECD, *La politique d'innovation en France*, Paris: 1986, pp. 294-300.

32 Dirk de Vos, *Governments and Microelectronics: The European Experience*, Ottawa: Science Council of Canada Background Study No. 49, March 1983.

sixth century, the Franks stopped the tide of invading Saracens.) Yet electronics has been the object of French government attention since the 1960s. The "intolerable" domination of IBM in the computer sector, culminating in the commercial failure of the technically successful French computer firm Compagnie Bull in 1965, was to be challenged by the 1966-1980 long-term Plan Calcul and the associated Plan des Composants. While financial support through the sector-wide actions concertées was still available, the main thrust of this "grand programme" was the creation of a "national champion" as an instrument of computer policy.[33]

The government caused the formation of a private company, the Compagnie Internationale de l'Informatique, CII, as a joint venture of three French electrical companies and contributed a grant of US$80 million as well as preferential treatment in government procurement. CII was to develop a comprehensive (small to large) line of computers and software. It is estimated that by the time it was obvious that CII failed in its mission, the government had expended $125 million on it. In 1971 a European consortium was formed under the title of UNIDATA and CII participated in this venture to build large computers. By 1975 UNIDATA was abandoned, at a cost to the French government of approximately $180 million.

In the meantime the failed Compagnie Bull was bought by American General Electric. When GE abandoned the computer business, it sold its interest in GE-Bull to Honeywell, which became Honeywell-Bull, or H-B. In 1976 after the collapse of UNIDATA, the French government persuaded H-B to merge with CII. The new computer company, CII-Honeywell Bull or CIIHB, was controlled from the outset by the French government (53 percent to a higher, but unknown proportion later on, of participation in the capital of the firm). Between 1975 and 1980, the government subsidized CIIHB to the tune of over $500 million, not including state-guaranteed procurement of about a $1 billion's worth of equipment over four years. While holding a leading position in revenue among non-American and non-Japanese computer firms,

---

33   Kristian S. Palda, *The Science Council's Weakest Link*, Vancouver: The Fraser Institute, 1979.

and a fair share of European markets, this company continued as a large drain on the French treasury in the form of support to innovation and direct subsidies (a substantial part of the 1982 informatique-microélectronique program budgeted at F650 million,[34] and F1.5 billion, respectively).

By the end of 1982 the French socialist government, elected the year before, effectively nationalized H-B, renamed it Machines Bull and invested, up to 1986, about F7 billion in it.[35] In the words of *The Economist*, the money allowed the firm to put off difficult decisions, like the move towards open systems and cutting surplus staff.[36] After Bull announced record losses of F3.3 billion in 1991, another rescue was mounted and the firm given an injection of F6.7 billion. This aroused strong misgivings in the European Common Market Competition Commission when the British computer firm ICL complained. After much hesitation, that subsidy was sanctioned. At the end of January 1992 the French government announced that it would allow the ancient adversary IBM to take a 5.7 percent share in Bull and share some technological advances with it. This share may prove larger, since the originally agreed price was F500 million, but later reports indicate F1.5 billion was paid.[37] Computers are, of course, but one part of the filière électronique for which the Ninth Plan scheduled a block appropriation of F140 billion over its five-year horizon.[38]

While it is difficult, and perhaps even inadvisable, to distinguish between innovation and industrial policy in France, the conclusion that

---

34   Projet de loi de finances pour 1982, Document annexe: la recherche, Paris: Imprimerie nationale, 1981, p. 28.

35   Jean Cheval, "New Trends in Industrial Performance and Industrial Policy in France," in Christopher Saunders, *Industrial Policies and Structural Change*, New York: Macmillan Press, 1987.

36   *The Economist*, April 20, 1991, p. 64.

37   *Globe and Mail*, January 29, 1992, p. B4; *The Economist*, June 22, 1992, p. 81; *Radio France Internationale*, June 23, 1992, 8 a.m.; *Globe and Mail*, July 3, 1992, p. B2.

38   *Globe and Mail*, January 27, 1983, p. B1.

French policy-makers still commit the bulk of their innovation support to the *grands programmes* that have not paid their way in the past is unavoidable. De Vos, who is an advocate of state intervention in this area, is compelled to admit that the government's incursion into the field of electronics and computers failed to wrench back control of domestic markets from foreign firms, failed to supply critical inputs from state-controlled sources, and failed to assure the necessary markets, both at home and abroad. In general, he states, it is hard to find any informed commentator who is not more conscious of failure than of success. In Zysman's opinion, the principal weakness of the "national champion" policy is that it creates firms whose basic skill must be to draw on the state's resources rather than to deliver market performance. In a world-wide competitive market setting, the dilemma is then that the protection and support required to produce specific products of interest to the state in fact weakens the firms that have been chosen to be the long-term instruments of state policy.[39]

It is inevitable that if the government takes sufficient risks with taxpayer money some of its gambles will pay off and that there will be benefits along the way. Spending substantial sums on the computer and components programs was bound to procure a respectable presence in certain markets and help strengthen the background skills and experience of French industry. It can even be argued that some state-sponsored software companies like Logabax have become both technically successful and commercially adequate after ending up in the private sector.

On the whole, in the electronics sector France has been very successful in telecommunications, where a well-defined, centralized purchasing, investment, and research policy has created what is probably the most advanced telecom system in the world. It has been equally unsuccessfull in computers and consumer electronics, where the competitive arena is unsuitable to state intervention. The latest example of this is the government plan to effect a merger between the ailing national consumer electronics group Thomson with other state companies. It is

---

39   John Zysman, *Political Strategies for Industrial Order: State, Market and Indus-try in France*, Berkeley: University of California Press, 1977.

an innovative idea of the fertile minds of the French bureaucracy: merge loss-making state-owned enterprises with profitable state monopolies—in this instance the nuclear supplier to Electricité de France, called CEA—and in this way avoid GATT and Common Market rules against trade in subsidized products. A similar deal meant to mask the weaknesses of the aerospace program was the taking by Crédit Lyonnais, a state-owned bank, of a 10 percent stake in Aérospatiale, the debt-laden government-controlled maker of Airbus, for F1.5 billion.[40]

The French government, undeterred by past failures in this sector, aims at establishing a European Electronics Community, a cooperative venture to defend European manufacturers against the "hegemony of Japanese and American firms."[41]

As one travels across France on the marvellous, straight-as-a-ruler secondary roads one wonders whether the engineers of the Napoleonic school of Ponts et chausseés could really, at the beginning of the nineteenth century, foresee the needs of twentieth century traffic, or whether they merely designed roads which appealed to their archimedic training. Straight roads were certainly unnecessary for coach travel. Similarly, a visitor nowadays to Versailles rarely stops to consider the 50,000 or so peasant victims of malaria and dysentery and the heavy tax burden that its construction exacted. One could also speculate that the Sun King's ostentation paid off in French material and intellectual as well as human capital exports to the rest of Europe.

Is it conceivable that if we discount over 150 years instead of over 15, the social rate of return turns positive and adequate? Or is it possible that the convenience of smart telephone cards, the Minitel video text terminal on the home phone, or the anxiety decreased by taking the TGV instead of the plane would turn the scales in favour of a positive social return as well? These things are hard to measure and remain largely unmeasured. Yet their immediate and narrow impact can be described

---

40 *Le Point*, December 21, 1991, pp. 61-62 and *Financial Times*, June 10, 1992, p. 15.

41 "France Places Bets on High Technology," *New York Times*, May 26, 1991, p. E5.

with enthusiasm, as is done so well in the special July 15, 1991 issue of *Time* magazine on France.

Back to Innovation Subsidies Elsewhere

In *Japan* the burden of taxpayer R&D support by tax concessions and subsidies is undoubtedly the lowest not only among the countries we compare here—Canada, Australia, United Kingdom, United States, Germany, and France—but also among all advanced industrial countries. At the same time, as we know, the GERD/GDP ratio is ahead of the most "intensive" spenders—Sweden, Germany, and the United States—as Table 45 shows. Table 44 indicates that in Japan in the mid '80s subsidies to private industry to stimulate precompetitive cooperative research amounted to roughly 40 percent of all government support. Thus, unlike most other OECD countries, the Japanese government channels very little support to individual firms. It is also frequently asserted that Japan's support to basic research, particularly that carried out in the universities, is niggardly, but is rapidly increasing.[42]

This, in a nutshell, is the *monetary* side of government support to innovation in Japanese industry. Of course, there is an interventionist side as well. While MITI, the fabled Ministry of International Trade and Industry, receives only between 10-15 percent of the government's share of total industrial R&D expenditures, recently it has been implementing a broad range of industrial technology policies, some through R&D projects in basic and leading edge fields. These are enumerated in detail below by citing directly from an OECD publication. It becomes apparent that such a wealth of projects cannot be financed from MITI's modest resources; industry itself pays for them under the direction of the persuasive ministry.

---

42   *The Economist*, August 11, 1990, p. 81; and Mowery and Rosenberg, *op.cit.*, p. 226.

Table 45: Selected International Comparisons of GERD, 1989

Country	GERD (US$ bil.)	GERD/ GDP (%)	Share Financed by Public Sector (%)	GERD per Capita (US$/person)
Japan	58.0	3.04	19.9	416
FRG	26.7	2.88	33.9	399
U.S.A.	130.3	2.83	48.2	553
U.K.[a]	17.0	2.20	36.5	298
France	19.0	2.32	49.9	312
Canada	6.7	1.33	44.0	246
Australia[a]	2.9	1.24	54.6	177

[a]1988.

Source: OECD, *Main Science and Technology Indicators*, 1991, No. 1.

The various projects include:

> The R&D Project on Basic Technologies for Future Industries, which promotes the development of such frontier technologies as new materials, biotechnology, and superconductivity; the Large-Scale Project, which brings together technological activities of both the government and private enterprises in areas such as the effective utilisation of marine bio-resources, and techniques for human sensory measurement; and the R&D Project on Medical and Welfare Equipment Technology, which is aimed at improving health and welfare through technological development.

> Japan has, moreover, been active in international research co-operation through such activities as shared research with foreign institutes, support for joint international research teams, invitations to foreign researchers to work in Japanese laboratories, and exchanges of research workers.[43]

---

43   OECD, *Industrial Policy in OECD Countries, Annual Review 1990* and *1991, ibid.*

At the same time, a number of projects is underway on environ-
ment-related industrial technologies while joint research programs
have been set up with foreign partners (the United States, Germany,
Brazil, etc.) to tackle problems of environment technology. The Japanese, wishing also to blunt the charges that for a long time they have been
free-riders on the bus of American and European technology, are very
keen on funding international precompetitive research projects. The one
that is the best funded and in full swing is described in Box 13.

Japan's industrial and industrial innovation policies have been
widely commented on and analyzed. This writer attempted to summarize some of this literature up to 1982 in a previous book. Since then
Kozo Yamamura has provided a masterful summary of such policies to
the mid '80s.[44] Most recently, Mowery and Rosenberg's reflections on
this subject seem to be particularly pertinent.[45] They opine that Japan's
"historic focus of policy on rapid absorption and intranational diffusion
of scientific and technological advances from foreign sources must be
supplemented by efforts to develop more of these advances indige-
nously, albeit in cooperation with scientists and engineers in other
industrial nations."[46] This is indeed what Japan now appears to be
attempting.

International Cooperative Research

The multitude of previous references to cooperative research leads to a
brief, separate section on *international* cooperation in innovation-creat-
ing or diffusing activities.

An international collaborative venture in research and development is, to paraphrase Mowery and Rosenberg, a collaboration between
firms or research organizations, including universities, that spans na-

---

44   Kozo Yamamura, "Caveat Emptor: The Industrial Policy of Japan," in Paul
     R. Krugman, ed., *Strategic Trade Policy and the New International Economics*,
     The MIT Press, Cambridge, MA: 1986, pp. 169-209.

45   Mowery and Rosenberg, *Technology and the Pursuit of Economic Growth*, *op.
     cit.*, pp. 219-237.

46   Mowery and Rosenberg, *op. cit.*, pp. 228.

Box 13: The Human Frontier Science Program (HFSP)

The Human Frontier Science Program (HFSP) aims at promoting, through international cooperation, basic research focused on the elucidation of the sophisticated and complex mechanisms of living organisms and to make the fullest possible utilisation of the research results for the benefit of all humankind.

The HFSP was originally proposed by the Japanese Government at the Venice Summit in 1987. Since fiscal year 1989, the Program has been supporting scientists around the world with their reserch activities that transcend national boundaries (through research grants, fellowships and workshops) in the context of international collaboration projects under the leadership of the Economic Summit member countries.

Research Areas

In order to make the best use of limited resources and to take into account the efficiency of the operation and management of activities such as the review procedures, certain priority research areas have been selected. Accordingly, the following two research areas are to be subsidized in the Human Frontier Science Program:

A. Basic research for the elucidation of brain functions. Priority research areas:
 • Perception and Cognition
 • Movement and behavior
 • Memory and Learning
 • Language and Thinking

B. Basic research for elucidating biological functions through molecular level approaches. Priority research areas:
 • Expression and Transfer of genetic information
 • Morphogenesis
 • Molecular recognition and response
 • Energy conversion

Research related to technologies that support basic research in the above two areas is also covered by the Program.

Activities and Size of the Program

A. Research Grants. Grants for basic research carried out by international joint research teams.

B. Fellowships. Fellowship support will be provided for researchers who wish to pursue research in foreign countries.

C. Workshops. Support will be provided for international workshop for the exchange of current information in the areas of research sponsored by HFSPO.

D. Size of Program. The number of research grants, fellowhips and workshops awarded each year will be determined by the organisation, taking into consideration the total funds allocated for each activity, the proposed budgets of applicants and the scientific quality of the applications received. In the second fiscal year (1990) of the Program, 32 research grants, 90 long-term fellowships and 3 workshops were awarded.

Source: OECD, *Industrial Policies in OECD Countries, Annual Review 1991*, p. 30.

tional boundaries, is not based on arm's-length market transactions, and includes substantial contributions by partners of capital, technology, or other assets.[47]

Box 10 described an "intranational" collaboration scheme or, as it is now fashionable to call it, a consortium. Some of the best-known *transnational* consortia are found in Europe. Among these are ESPRIT and EUREKA.

ESPRIT, the European Strategic Programme for Research in Information Technology was fully launched in 1983 as a response to Japanese challenges in microelectronics, software, office systems, computer-integrated manufacturing and information processing. Started on the advice of the twelve largest European electronics and telecommunications firms under the chairmanship of the Common Market commissioner for industry Vicomte Davignon,[48] it has reached in its Phase II (1988-92) planned expenditures of ECU 3.2 billion.[49] About half of this funding comes from the European Community and half from the participants, who are both firms and research or university institutes.

Under the "umbrella" strategic guidance of the Brussels Commission, ESPRIT is a collection of independent projects. These projects must be proposed by at least two independent firms or institutes from at least two nations, of which one must be an EC member nation. A group of outside consultants to the Commission evaluates the prospects. If accepted, the Commission's guidelines on information sharing and patent/royalty rights apply. In general the research undertaken concerns pre-competitive technologies, a stage at which marketability is not of primary concern as yet.

On the heels of ESPRIT, the French-proposed EUREKA program was initially planned as a response to the US SDI or Star Wars initiative. The French feared that SDI would cause a brain drain from Europe's defence and electronics industries. Instead of a centrally-funded EC

---

47   *op. cit.*, p. 242.

48   Margaret Sharp, "Europe: Collaboration in the High Technology Sectors," *Oxford Review of Economic Policy*, 1987, pp. 52-65.

49   Todd A. Watkins, "A Technological Communications Cost Model of R&D Consortia as Public Policy," *Research Policy*, 20 (1991), pp. 87-107.

scheme, however, EUREKA has evolved into a program funded by individual countries and their industrial firms/research institutes/universities with a very small secretariat in Brussels which acts mainly as a marriage broker.

Watkins, whose article probably represents the most accessible recent report and analysis of ESPRIT and EUREKA states that the latter has no strategic priorities with respect to technologies: supported are projects, for instance, in fields as varied as biotechnology, lasers and transportation.[50] EUREKA concentrates on projects further down the road to development than ESPRIT. Once the Brussels secretariat gives its badge of approval to a project—the conditions are much the same as under ESPRIT—individual countries' governments are free to fund their own institutions participating in it; no funding "crosses borders." As of late 1989 close to 300 projects were running.

JESSI, the Joint European Submicron SIlicon project, started in 1989 under EUREKA appears to be the largest single cooperative project here, with ECU 3.5 billion pledged to its seven-year program.

The cornucopia of varied large and small projects, with big and small participants, would appear to have alleviated the fear that these European Community sponsored programs were to become, in the words of Sharp, merely transfers of the old national champion programs to the European level, creating in their place European champions who in their turn require further subsidies and protection.[51]

Programs that support research consortia are usually recommended on the grounds that they help to overcome market failure: they enable firms to spread risks and achieve the critical minimum efficient scale to lower costs. Research cooperation may also alleviate the appropriability problem. If it lowers the costs of innovation to the member firms and makes the outcomes accessible to all of them, investment in R&D may increase as the inappropriability disincentive diminishes.

Yet overcoming market failure seems not the whole story. Watkins assembles arguments that there are substantial transaction costs in

---

50   *ibid.*

51   Sharp, *op. cit.*

communicating technical knowledge and that the human contacts and the learning of others' working methods implied in the inter-firm or inter-institution cooperation are very important in lowering such costs.[52]

Watkins also argues, using as a theoretical background Buchanan's theory of clubs, that the so-called *à la carte* approach of EUREKA, in which firms and governments are not faced with all-or-nothing choices and financing, increased the political acceptability of innovative support. Given that cross-border cooperation of individual firms/institutes is desirable, the political arrangement of umbrella-sponsorship of voluntary *ad hoc* associations succeeded in splitting up "one problematic negotiation into a series of more tractable pieces involving groups with more homogeneous interests."[53]

We should not, however, be bowled over by pro-intervention arguments, whether they issue from the mouths of Harvard professors or Eurocrats. *In puncto* JESSI, for instance, in 1990 work had barely begun on a substantial part of its project than it ran into difficulty.

The plan was to lay the foundation to X-ray chip lithography and then, with its help, to develop a 64M-DRAM (dynamic random access memory) and a 16M-SRAM (static random access memory). But very soon Philips withdrew from the production of SRAM's and so this portion of the project fell through. It also became apparent that optical lithography, the current production technology, did not exhaust all possibilities of further technological improvement. This decreased the interest of European semiconductor manufacturers in X-ray lithography considerably.

The third blow was the decision by Siemens to develop 64M-DRAMs in bilateral cooperation with IBM rather than in the JESSI framework. Writing in December 1991, two German economists state that of the JESSI concept there remains barely the name—and the pledged DM 8 billion that must be spent.[54]

---

52   Watkins, *op. cit.*

53   *ibid*, p. 103.

54   Georg Bletschacher and Henning Klodt, *Braucht Europa eine neue Industriepolitik?* Kiel Discussion Paper 177, December 1991.

Concluding Remarks

This chapter outlined a variety of government taxation, subsidy and other types of support to the creation and diffusion of innovation in Australia, France, Germany, Japan, the United Kingdom, and the United States.[55] The generosity and variety of taxpayer aid given to innovative activities in these advanced industrial countries is impressive. Perhaps less impressive is the extent of efforts invested in evaluating the results of all this support and in assessing the waste of taxpayer contributions.

Yet there are two positive reflections to be made. Governments have fashioned the innovation component of their industrial policies with a view to appropriating the benefits of research for their nations.[56] But these same governments have paradoxically encouraged the formation of transnational research consortia which actually enhance the dissemination of technological knowledge. Unwittingly, perhaps, these governments seem to acknowledge tacitly the ever-increasing globalization and intertwining of industrial arts. This bodes well for other areas of the economy that are afflicted by artificial barriers and rent-seeking protectionism.

And then one should also remember that of all the squandered resources in contra-economic subsidization, those expended on an increase in knowledge—rather than on an increase in wine lakes, butter mountains or regulators—seem to be the least wasted.

---

55 We have not really mentioned patent policies, because they are similar in all these countries and in Canada (since the reform of the pharmaceutical patent act in 1987). For another look at government intervention in the innovation area see OECD's *Science and Technology Policy*, Paris: 1992.

56 Mowery and Rosenberg, *op. cit.*, p. 289.

Chapter 8

Implications

The arguments put forward by proponents of increased government support for R&D in Canada have now been critically examined and their theoretical and empirical validity found questionable.[1]

Yehuda Kotowitz

Introduction

In a recent article Professor Paul Krugman of the Massachusetts Institute of Technology, one of the founding fathers of strategic trade policy analysis, offered this valuable remark:

> ... an intellectually respectable argument can be made for government policies to create or preserve [comparative] advantage. The fact that an argument is intellectually respectable does not mean that it is right. Concerns over competitiveness that are valid in principle can be and have been misused or abused in practice. Competitiveness is both a subtler and more problematic issue than is generally understood.[2]

---

1 *Positive Industrial Policy: The Implications for R&D*, Toronto: Ontario Economic Council, 1985, p. 33.

2 Paul A. Krugman, "Myths and Realities of U.S. Competitiveness," *Science,*

Krugman's statement summarizes neatly the broad theme of the present volume. Too often in Canada intellectually feasible arguments for government intervention in high technology have been put forward with the force of truth. Those espousing intervention belong typically to a self-interested hodge-podge of industry lobbyists, compliant bureaucrats, and politicians mesmerized by visions of unlimited growth through R&D. But as any economist who has played with theoretical models will claim, feasible arguments can be manufactured for any policy position. It is more difficult to fabricate support for a position, however, if one is obliged to pay attention to evidence.

The evidence which this book has presented does not support the case for intervention on the present scale and scope. The purpose of this final chapter is to restate when intervention is reasonable and when it is not, and to bring attention to how various interested groups have blown the merits of intervention out of proportion and have proposed questionable solutions.

One of the most persistent notions which has developed among policy makers is that much of the R&D a business performs spills over to other firms. Firms, it is argued, will "free-ride" on the information provided and undertake too little R&D as a result. Government must subsidize R&D to set things straight. The problem with this argument is one of missing facts. In practice, businesses that are not active in R&D cannot make much use of information spillovers. Even if they could, however, it is not clear that government subsidies to R&D would lead to greater innovativeness. There is no evidence which suggests this to be the case.

Where arguments about spillovers fail to be convincing, R&D subsidy advocates have pointed to Canada's need to be competitive. Only an early investment in the technologies of tomorrow can assure our future international standing. Part of the solution, we are told, is to support investments in technologically intensive sectors. This is another example of reasoning severed from the facts and rooted in a firm misunderstanding of what competitiveness means and what R&D achieves.

November 8, 1991, pp. 811-815.

To understand the nuances of this debate and how certain groups have twisted it to their advantage it is necessary to examine both sides of the story and the evidence accumulated to date. In the high technology debate, it is also important to ask *cui bono*—who profits? In many cases the answer is "no one." Some policies are simply misguided antidotes for which politicians reach in the hope of curing economic ills they themselves have wrought on the economy. Other policies have clearly identifiable beneficiaries, most often highly skilled and trained scientific workers, or ambitious investors. This chapter summarizes the arguments, pinpoints the beneficiaries, and repeats the conclusion echoed throughout this book that Canada is not deficient in its support of R&D.

Innovation Policies: For and Against

Let us start with Krugman's statement that competitiveness (we may easily substitute innovativeness here) is a more problematic issue than is generally understood. The illustration of that assertion is known as the Schumpeterian dilemma.[3] The stimulus to innovation can also be an obstacle to diffusion: a firm innovates the more, the more it can appropriate the benefits of that innovation for itself. One way to appropriate those benefits is to exclude competitors from the market.

A prominent example of this dilemma is the multiphase wrangling over patent protection in the Canadian pharmaceutical industry, already mentioned in chapter 6. The latest development in this area is a federal bill passed in 1992 which adds another three years of effective patent protection to patented, brand name drugs against low-cost generic versions. Under this bill all vestiges of compulsory licensing of patented drugs to generic manufacturers were eliminated.

The bill was welcomed by the patented medicines manufacturers with the pledge of an additional $400 million in capital and research spending over 1993-97, additional to the $1.1 billion already pledged

---

3   In his pathbreaking *Theory of Economic Development* (Harvard College, 1934), Joseph Schumpeter made a strong case that the promise of a temporary market monopoly is the driving force behind a firm's innovative activity.

and partially carried out as a consequence of the first legislative enhancement of patent protection in 1987. It was vigorously opposed by the generic competitors who predicted dire consequences for their own investment and employment. And it was generally deplored, with the exception of Quebec which stands to reap most of the R&D employment increases, by Canada's provincial health ministers fearing increased drug prices.[4]

This Canadian case illustrates perhaps better than any other the conclusions reached in an influential paper commissioned by the OECD.[5] Namely, that arguments in favour of government intervention should rest on detailed qualitative descriptions of the marketplace and often on specific parameters that describe conditions there. These arguments do not apply across the board; the nature of the problem makes case-by-case analysis unavoidable. In other words, there are no satisfactory, all-embracing grounds on which to base government intervention.

The argument used most frequently for taxpayer support of innovation is that research resulting in innovation leads to the creation of knowledge. Knowledge has public good characteristics. It is non-exhaustive in use, unlike most other goods and services, since the use of information by one agent does not eliminate its availability to another one. Knowledge is also imperfectly excludable. Agents who have incurred costs in creating it often find it difficult to protect their intellectual property rights, despite secretiveness or patent rights. Free-riding by imitators can frequently occur. Such public good characteristics of new knowledge lead to leakages or spillovers of valuable information from innovators to competitors and throughout society.

The presence of spillovers will discourage private agents from undertaking any generation of new technology beyond the benefits that can be captured or appropriated. It is believed that R&D will be underprovided by the market system to the extent that spillovers take

---

4   *The Globe and Mail*, various pages in the June 23 and 24, 1992 editions.

5   Gene M. Grossman, "Promoting New Industrial Activities: A Survey of Recent Arguments and Evidence," *OECD Economic Studies*, No. 14, Spring 1990.

place, while dissemination will be sub-optimal when spillovers are prevented.[6]

The existence of spillovers, despite some reservations among certain researchers, appears to have been empirically established. By now we would thus seem to have substantial evidence of a gap between private and social returns—evidence of inappropriability. But is this *prima facie* reason for taxpayer support of private investments in innovation?

We also have a reasonable theoretical framework for procedures that would guide the subsidizer in determining the need for a subsidy to counteract the inappropriable returns and in estimating its optimal size. A simple formula, proposed in 1984 by the Canadian economist Tarasofsky and explained in chapter 6 is one promising formulation that looks applicable. A more complex approach, proposed by Grossman in 1990, carries the name of "surplus analysis."[7] Both economists stress that the case that emerges for government subsidy must be tempered to reflect the "excess burden" or deadweight cost of raising the necessary funds.

The core of the argument in this book lies in the debate about industrial policy—subsidization in the most general sense—toward innovative activity.[8] If, on the basis of empirical evidence, we find that social returns to innovation exceed private ones, is this not a necessary precondition for the policy established? If we have good theory to guide the policy's efforts, is this not a sufficient condition established for subsidization?

No, not if the empirical evidence on inappropriability does not point unambiguously in the direction of R&D underinvestment; not if the theoretical prescriptions for carrying out optimal subsidy allocation are not capable of being successfully applied.

As Kotowitz stated in his Ontario Economic Council monograph:

---

6   Grossman, *op.cit.*, p. 106.

7   *Op.cit.*, p. 90.

8   Subsidization in the form of grants, tax advantages, regulatory or trade protection, and purchasing preferment.

Two questions must be answered affirmatively before government support for industry can be justified from an economic point of view. First, do impediments exist that seriously restrict the ability of the private market to take advantage of all opportunities and therefore to realize the allocation of resources among industries and activities that is optimal from a collective point of view? Second, can government do better?[9]

The first question addresses the seriousness or otherwise of inappropriability or spillovers, and perhaps of risk and insufficient scale. Let us consider inappropriability, a consequence of non-excludability.

Bernstein in Canada (see chapter 2) and others in the United States uncovered a substantial occurrence of spillovers. Why is it then that we see all around us research being carried out privately, new processes employed in manufacturing, distribution, and services, new products streaming onto the market? This must mean that not all new knowledge is easily transferrable (non-excludable) or that it is imperishable (non-exhaustive). In other words, new technological knowledge is not always afflicted with the stigma of a public good, with the consequence that it is not appropriable.

There is, of course, patent protection and commercial secrecy.[10] But with respect to leaks or spillovers there is a growing awareness that for a business to absorb spilling technology from competitors or adjacent sectors it must itself engage in substantial R&D activity. Cohen and Levinthal, in their influential 1989 article previously cited in chapter 2, show that a firm's absorptive capacity is a positive function of its own R&D effort.[11] It is therefore often the case that R&D rivalry among

---

9   Kotowitz, *op.cit.*, p. 3.

10   Firms mitigate the danger of inappropriability with the following instruments: commercial secrecy, rapid diffusion of one's own product to preempt the market, rapid obsolescence of one's own research, joint R&D ventures, the use of government lab results, patents. See Claude Crampes, "R&D, Appropriability and Competition," paper presented at the international seminar on technological appropriation organized by INSEE, Paris, June 9-10, 1992.

11   W.M. Cohen and D.A. Levinthal, "Innovation and Learning: The Two Faces of R&D," *Economic Journal*, 1989, pp. 543-565.

competitors, or innovative absorption of spilling knowledge may alleviate or even eliminate the consequences of suboptimal spending by "original" innovators. Anyone familiar with the research behaviour of leading pharmaceutical firms will discern the validity of this assertion.

The second alleged public good characteristic of new technology is its non-exhaustiveness: the use of it by one firm does not limit its consumability by another. While in strict theory this is undoubtedly so, the practical picture looks different. Very often the advance in knowledge is a perishable commodity, overtaken soon by yet future advances. In the semiconductor industry, for instance, the average lifecycle of a chip in the 1980s was around 3 years.[12] The manufacturer-innovator of a new version, such as that of a 256K-DRAM over a 64K-DRAM, raced down—with the help of process R&D—the learning curve so fast that he left his competition well behind. Yet, for each new chip type the manufacturing process had to be substantially re-arranged again, leaving the previous innovator with but little advantage. Thus what counts is speed and ever-renewed research, not questions of leisurely imitation.

We can now return to the first question posed by Kotowitz: do impediments exist that hamper an optimal allocation of resources to innovative activity? Yes, of course they do, but close scrutiny reveals a shrinking number of them. This undermines seriously the appropriability paradigm which is based on the notion of an essentially costless transfer of knowledge—a paradigm that used to provide, in the words of Mowery, "a simple, elegant prescription of considerable generality for the policymaker. Data on benefits, costs, private, and social returns are compact and easily understood inputs for policy analysis."[13]

What is left for government intervention is a nuanced, tricky analysis which must first uncover industries that suffer from inappropriabil-

---

12   G. Bletschacher and Henning Klodt, "Braucht Europa eine neue Industriepolitik?" *Kieler Discussion Paper 177*, Institut fur Weltwirtschaft, December 1991. See also David T. Methé, "The Influence of Technological and Demand Factors on Firm Size and Industrial Structure in the DRAM Market, 1973-1988," *Research Policy*, 1992, pp. 13-25.

13   David C. Mowery, "Economic Theory and Government Technology Policy," *Policy Sciences*, 1983, pp. 27-43.

ity and then find out to what extent the knowledge spilling inside or out of this industry depresses innovative or R&D activity of both the spillers and the receptor sectors or companies. This is a tall order, only partially fulfilled by the Tarasofsky prescription. But, as we have seen, Tarasofsky could not find any indication of adherence even to the simple principles contained in his advice in a number of federal subsidy programs for innovation. In addition, the analysis required to uncover deserving individual recipients of subsidy is really quite infeasible in a tax concession context. And, of course, taxes are now a predominant vehicle of subsidy support to R&D in Canada and, as was pointed out in the previous chapter, in a number of other countries.

The second question posed by Kotowitz is starkly simpler. If market failure depresses innovative activity in the private sector, can government do better than the market? The answer here rests on the efficiency of the politico-bureaucratic delivery of market failure remedies.

The first condition for efficient delivery is the availability of sound economic guidance for the analysis of who the deserving recipients should be. We have noted the emerging findings that while spillovers (and so inappropriability) abound, this does not necessarily mean that overall innovative activity is thereby depressed. Therefore, on current knowledge, as opposed to our certainties a dozen years ago, economists cannot provide firm practical guidance to industrial policy subsidization of private innovative activity. And so, as Grossman points out:

> It will of course be difficult for the policy analyst to identify the deserving innovations and to delimit the period of government support to the time where substantial externalities are being generated.[14]

The second condition is the practical one of how to deliver support to firms and industries. Box 3 illustrated how a government assistance program, the Industrial Regional Development Program (1983-88), suffered in its application by the lack of a clear economic rationale. The Box 10 example of Sematech showed how the emphasis on precompetitiveness precluded a U.S. taxpayer-financed venture in microchips to reach desired learning effects. The end of the last chapter described how

---

14   *Op.cit.*, p. 119.

changing competitive and technological circumstances wrecked the original goals of JESSI. Yet this did not result in refunding the planned outlays to Common Market taxpayers.

These are all instances of the politico-bureaucratic process of allocation going haywire. Perhaps even more frequent is the loss of taxpayer money on ventures that ultimately do not pan out for technological rather than "administrative" reasons. The most glaring Canadian example is turning out to be the CANDU reactor, though the final word is not in. For Japan's fabled MITI it is likely to be HDTV, the high definition television, in which all bets were placed on conventional analog technology. The United States government resisted a determined push by a much-lobbied Congress to subsidize joint research to speed up HDTV development.[15] Now it is clear that digital rather than analog transmission technology is the way to go. Unsubsidized U.S. firms surged ahead of the Japanese and are awaiting U.S. Federal Communications Commission approval for national broadcasting in 1997.[16]

It is perfectly natural that some technology forecasts go wrong. What is less natural is that it should be the taxpayer rather than the entrepreneur who bears the cost. In the CANDU case General Electric was willing to carry on with manufacture. In the HDTV case some American enterprises, such as General Instrument, were able to do without subsidy. This is the usual pattern with government subsidization of innovation. It breaks the rule: do not support A if B is carrying on without subsidy—a rule mentioned in chapter 6. Of course it goes against the grain of strategic government planners who insist that we must be there first and pick winners.

In closing this section let us repeat its main points. A theoretical case can be made for taxpayer support to innovative ventures, bolstered by empirical evidence of spilling, or inappropriable knowledge passing from "imitators" to others. However, recent findings about the costs of spillover absorption and about rivalry among innovators undermine the claim that inappropriability on the part of innovators will necessar-

---

15   Cynthia A. *Beltz, High-Tech Maneuver: Industrial Policy Lessons of HDTV*, Lanham, MD: University Press of America, 1992.

16   "Japan Refocusing HDTV," *New York Times*, July 3, 1992, p. B4.

ily depress the overall level of innovative activity. This makes the enhancement of innovativeness, whether by grants or tax incentives, very difficult to carry out in a "socially optimal" manner. Furthermore, the practical question of delivering aid to innovation often falters expensively, either for reasons of politico-bureaucratic complexity or because the subsidized technology is changing rapidly. All of this urges extreme caution in allocating taxpayer assistance to innovation.

Thus far we have addressed some "objective" impediments to an effective public stimulation of private innovativeness. "Subjective" impediments deserve a few paragraphs.

The Private Motives Behind Public Support

As is frequently the case with subsidy programs, whether to agriculture, fisheries, or industry, support of innovation or R&D is often the result of a convergence of private interests operating against a background of public interest.

Typically, there are three groups determining the amount and distribution of taxpayer assistance to innovative efforts: the special interest lobby, the political entrepreneurs and the administering bureaucracy.

Nowadays in Canada, as in most industrial economies, it is commonplace for firms or their associations to seek favours—rents in the parlance of economists—from governments. As a *Globe and Mail* columnist put it, "in industry after industry, and in all provinces, corporations are lining up for grants, loans and concessions. And politicians are obliging."[17] Research intensive industries have the most to gain from favours to innovative activities and most of them, by dint of the nature of their work, happen to be in a supplier position to the government anyway. As Pavitt stated some time ago, large-scale involvement of governments in high technology has led to the creation of large groups of influential and able scientists and engineers. The result of their work is no longer subject to the ultimate sanction of the markets, and they

---

17   Terence Corcoran, "Will Bob Rae Hand Out Another Free Lunch?" *Globe and Mail*, May 27, 1992, p. B2.

themselves create the demand which only they can satisfy. Society has in general failed to establish control over the ventures they propose and carry out.[18]

Political entrepreneurs find it attractive to supply innovative policies because such policies seem to enjoy favourable publicity and broad political support. Arguments about positive externalities of research and the need to keep abreast of the efforts of foreign governments seem to be widely accepted by the electorate. At the same time, the degree of ignorance and the costs of getting information are especially high for this kind of policy for the average elector. As Streit puts it, the asymmetric information in favour of the politician provides him with opportunities to serve special interest groups.[19]

Furthermore, as Streit points out, the risk of being blamed for any policy failure is very low. Given the desirability of the policy in general, the political action becomes more important than its technical, let alone economic outcome, especially since the influence of the policy on such outcomes is particularly difficult to judge. Finally, as was stated in chapter 1, the impossibility of macroeconomic manipulation in the face of fiscal bankruptcy seems to underscore the political attractiveness of support to "hi-tech," where something still apparently "can be done" to inject growth into the economy.

The interests of a budget-maximizing bureaucracy are clearly allied in the instance of technology policies with those of the vote-maximizing political entrepreneurs. This applies above all to grant programs which necessitate the processing, approval, and supervision of individual projects.[20] Apart from the bureaucratic input that this requires, one should not underestimate the various fees flowing to outside consultants and the consequent return of favours to project administrators. In Germany, for instance, Streit mentions that between 1972 and 1980 the

---

18 Keith Pavitt, "Government Support for Industrial Research and Development in France: Theory and Practice," *Minerva*, Autumn 1976, pp. 330-54.

19 Manfred E. Streit, "Industrial Policies for Technological Change: The Case of West Germany," in Christopher Saunders (ed.), *Industrial Policies and Structural Change*, New York: Macmillan Press, 1987, pp. 129-142.

20 See, for instance, those listed in tables 3 and 29.

total number of projects dealt with by the Ministry of Research more than trebled, from 2,079 to 7,000, whereas its total expenditures for R&D only doubled.[21]

In brief, technology policy, and in particular R&D subsidization, present a fertile opportunity for the extraction of rents from taxpayers.[22] It is very costly for citizens to inform themselves about the issues involved (asymmetric information) and at the same time easy for them to be attracted by the glamour of high technology. This presents a suitable entry point to rent-seeking interest groups and finds active support from budget-making politicians and enabling bureaucracy. And even if a large-scale financial scandal ensues, as is the case of the $2 billion-plus SRTC boondoggle, who remembers those responsible? How many Canadians recall a Liberal finance minister by the name of Marc Lalonde who introduced the legislation? The obscurity in which high-technology lobbies work, and the lack of blame politicians and bureacrats can expect to bear for failures of massive scale are perhaps why Grossman writes that

> The potential societal gains from an activist policy can easily be sacrificed if opportunities for wasteful rent seeking are created or if the criteria for selection become the political clout of the applicant rather than the economic merits of the case.[23]

Competitiveness an Excuse for Subsidies?

The recent awakening of the public to Canada's need to be competitive has renewed the call by the high-technology lobby for government support to R&D. Whether these calls are warranted, however, is in doubt. To see this it is important to ask how competitiveness is defined, what is Canada's situation, to what extent is technology responsible for

---

21   Streit, *op. cit.*, pp. 133.

22   Linda Cohen and Roger Noll, *The Technology Pork Barrel*, Washington: The Brookings Institute, 1991.

23   Grossman, *op.cit.*, p. 119.

competitiveness—and what remedies are proposed against competitive decline.

As was already indicated, the definition of competitiveness includes, but is not limited to (foreign) trade performance. The happiest one-long-page definition is to be found in an exceptionally well written 1992 OECD publication.[24] That is too long for our purposes here, so we opt for presenting a definition by the (U.S.) President's Competitiveness Policy Council:

> What is Competitiveness?
>
> The Council's definition focuses on four criteria. First, U.S. goods and services should be of comparable quality and price to those produced abroad. Second, the sale of these goods and services should generate sufficient U.S. economic growth to increase the incomes of all Americans. Third, investment in the labor and capital necessary to produce these goods and services should be financed through national savings so that the nation does not continue to run up large amounts of debt as in the 1980s. Fourth, to remain competitive over the long run, the nation should make adequate provisions to meet all these tests on a continuing basis.[25]

There has recently been a flood of reports and books on the subject of competitiveness including, in Canada, the Porter report and a "consensus" report of the now defunct Economic Council of Canada.[26] Both of these come to the inevitable conclusion that not all is well with Canadian competitiveness.

The Economic Council report pays considerable attention to the decline in the rate of growth in real GDP per capita for 1962 to 1990. It states that a major cause of this decline was the slowing of TPF, total factor productivity. Total factor productivity includes both capital and

---

24 OECD, *Technology and the Economy*, Paris: 1992, p. 237.

25 Competitiveness Policy Council, *Building a Competitive America*, Washington: March 1, 1992, 2. A very useful short bibliography is included.

26 Michael E. Porter and Monitor Company, *Canada at the Crossroads*, Ottawa: Business Council on National Issues, October 1991; and Economic Council of Canada, *Pulling Together*, Ottawa, 1992.

labour productivities and its change, while influenced by technological innovation (see Table 1), depends on other factors as well.[27] The Council estimates that 16 percent of the growth in labour productivity in the past was accounted for by R&D expenditures.

Porter's report looked also at individual industries rather than just the broadest picture. It speaks of the comfortable insularity of the old order and finds fault, among other things, with Canada's innovativeness. The interesting thing in both reports is that precise suggestions for direct governmental stimulation of industry are lacking. This can be interpreted as indicating either that public policy toward technological innovation is not deficient or technology is not the prime culprit.

Is technology but a secondary player in the decline of competitiveness? The Competitiveness Policy Council report gives it just three pages out of 37. *The Economist*, reporting on June 27, 1992 on the then as yet unpublished paper by Porter addressed to U.S. industrial problems, lists 31 recommendations that the scholar makes to policy-makers, institutional investors, and companies.[28] Only one refers to technology: "improve corporate investment incentives—provide investment incentives for R&D and training."

A suspicion that government innovation policy in Canada is not deficient can be read into the following quote from the Economic Council's *Pulling Together*:

> In other words, through the tax system governments pay for more R&D by industry than industry itself does.
>
> This tax treatment of R&D expenditures by the private sector may be the most generous in the world. The problem of low R&D expenditures by industry clearly does not lie in the overall amount of government support for this activity. Therefore, it must lie elsewhere—either in a weak commitment to research on the part of the industrial sector or in an inadequate composition of the government funding. In any event, if management

---

27 For instance the shift of employment from manufacturing to service sectors, where a change in productivity is much harder to detect.

28 Michael Porter, "Capital Choices: Changing the Ways America Invests in Industry." Harvard Business School is the only "identifier" in *The Economist*, June 27, 1992, pages 89-92 article.

in Canadian industry has no commitment to research, then it is unlikely that government can alter its behaviour by offering incentives.[29]

And so we have arrived at the nub of the basic argument as far as policy toward innovation is concerned, through a brief stop at the competitiveness issue. Here is OECD's opinion:

> The broader lesson to be drawn from the past fifteen years' experience is that many Member countries have tended to overrate the risks and the costs of such market failures and to underestimate those associated with "government failures." These government failures have often resulted from an inclination to underrate the stringent conditions required. These requirements are that the objectives be clear and realistic, that the means match ends, that the measures be compatible with these policies and that the *seed sown by government falls on fertile ground* [my italics]. On this latter fundamental point, government programmes have often been hampered by a lack of receptiveness and/or capacity *to respond to incentives*—factors which hinge on the quality of the macroeconomic, financial and educational environment of industry.[30]

We concur with the OECD and with a growing chorus of other economists[31] and enlightened politicians, that factors other than just research and development influence innovation, and that prominent among them are "business conditions" described in the model suggested in Figure 1. Even the federal government agrees:

> The GERD:GDP ratio makes an assumption that a direct relationship exists between the amount of R&D performed and the productivity of a nation. There is, however, much more involved in improving a nation's competitiveness. Access to markets and capital, the availability of skilled human resources, the capacity of Canadian industry to adapt and adopt technology

---

29   *Pulling Together, op.cit.*, p. 50.

30   OECD, *Structural Adjustment and Economic Performance*, Paris: 1987, p. 234.

31   Luc Soete, "Technology and Economy in a Changing World," background paper prepared for the Conference on technology and the global economy, Montreal, February 3-6, 1991.

and business framework policies, all have an important impact on productivity.[32]

What are the "business framework" policies that make investment in innovativeness attractive? And why are they so seemingly difficult to put in place, in Canada or elsewhere?

In the first place, they are the macroeconomic policies that minimize the uncertainties surrounding the choices of industrial firms and are sufficiently neutral in their microeconomic effects as not to distort these choices. Everybody agrees that a stable currency and low interest rates as well as reasonable tax rates are the important ingredients favouring investments of any kind, including those necessary to invention and adoption.

In addition, it is desirable to have an adequate degree of competition as a stimulus to excellence and as a guarantor of satisfactory allocation of human and capital resources. In a small economy such as Canada's, free trade is an automatic promoter of such competition. It is of interest to cite on this point the Business and Industry Advisory Committee to the OECD:

> BIAC concludes that, from the perspective of industry, the most effective means for governments to promote technology and its contribution to economic and social objectives is to create the conditions for increased competition. In this regard, the most effective actions that can be taken by governments would be to redouble efforts to liberalize markets and industries at both the national and international level. This will provide the boost for competition which spurs firms to create and use technologies effectively.[33]

A leg-up to competition by governments includes a restraint from bail-outs of any form. Thus Porter says:

---

32   Government of Canada, *Response to "Canada Must Compete,"* A report of the standing committee of the House of Commons on Industry, Science and Technology, Ottawa: Sessional Paper, May 13, 1991, p. 6.

33   BIAC, *Principles for the Promotion of Science, Technology and Industry—A View From the Industry*, Paris, February 23, 1990.

Canadian firms must re-evaluate their expectations of government and place different demands on government than in the past ... they should no longer look to government to provide traditional forms of assistance—subsidies, artificial cost structure, lax regulations, guaranteed procurement.[34]

And the Economic Council states:

It [the Council] also recommends that governments avoid bailing out firms through subsidies or import quotas when they have little prospect for longer-term efficiency.[35]

And the February 1991 Montreal International Policy Conference on technology and the global economy opines:

The shift in the respective roles of industry and governments *vis-à-vis* technological change also necessitates profound changes in attitude. In particular, governments must no longer be looked upon by the private sector as purveyors of lifebelts, whose main function is to keep certain firms or sectors afloat— particularly when they can be described as *strategic*—at difficult times or when markets are malfunctioning. Only on this condition will governments be fully able to facilitate innovation and competitiveness.[36]

The third, equally important ingredient of "framework" policies that contribute to the quality of business conditions is an education and training system able to supply—and retrain if necessary—the people to meet the technological demands of present-day production and innovation.

It is unnecessary to return to the discussion of the specific innovation and diffusion support policies that are generally recommended, for as we have seen, Canadian governments have provided an abundance of them, without—at least according to the numerous critics—apparent success. Rather, we shall return to the educational infrastructure that would seem to enhance innovation to show where at least one of the three "business framework" policies seems in straits.

---

34   Porter, *op. cit.*, p. 83.

35   *Au courant*, Vol. 13, No. 1, 1992, p. 10.

36   *Op. cit.*, p. 7.

Before we do so, let us address the question posed earlier: why are the chief ingredients of the framework policies so difficult to put in place? The basic answer, due mostly to debates in the public choice area of economics over the last quarter century, is that governments have become hostage to special group interests that suck the blood out of tax receipts. This does not leave enough money for government to provide the traditional fiscal and monetary framework and the infrastruture in which it may have comparative advantage.[37] Support for education is part of the infrastructure that appears to be crumbling in Canada in the face of other pressures, such as the financing of regional unemployment rigor mortis or of badly designed health care systems.

If there is unanimity among students of innovation and competitiveness, it is on the importance of enhancement of human capital by education and training. Thus, for instance, the astonishing success of post-war Japanese economic growth is laid at the door of the generally high educational level of the population which allowed the specialization in human capital—intensive production.[38] When human knowledge replaces machines and plants as the moving factor behind productivity change, and when the required knowledge is missing, output suffers however much finance and hardware are supplied.[39] Ergas, a foremost student of innovation policies, insists particularly on the role of education and training in technology diffusion:

> The flow of newly trained personnel into the active population allows the continuous upgrading of skills and capabilities. At the same time, the better educated the labor force is, the greater will be its capacity to adjust to sophisticated new techniques. Higher levels of education are also likely to make this capacity more widespread, both throughout industry and throughout the active population.
>
> Countries whose investment in human capital lacks depth or breadth may be among the pioneers in generating new technol-

---

37   Anne O. Krueger, "Government Failures in Development," *Journal of Economic Perspectives*, August 1990, p. 17.

38   Bletschacher and Klodt, *op.cit.*, p. 32.

39   OECD, *Technology and the Economy, op.cit.*, p. 150.

ogies, given a sufficiently strong scientific elite. But as far as using these technologies is concerned, they will be disadvantaged on two counts: an inadequate rate of expansion or replacement of the skill base at the margin and difficulties in adjusting the existing stock to the demands of technological change.[40]

Canada's Economic Council concurs:

Overcoming these (economic) hardships will require substantially increased investments in training, skill development, and mobility. Human-resource development policies must be considered the *cornerstone of a strategy of national competitiveness* [my italics].[41]

Despite all these admonishments governments, especially provincial governments who are responsible for education, prefer or are compelled to divert funds elsewhere. The prime victim is higher education, as Figure 22 shows. Figure 23 gives a "competitive" comparison with the United States. Financing is not, of course, the whole story. Regulation, too, is detrimental since, for instance, the provincial governments do not allow their universities to increase their tuition fees as they see fit. This prevents the emergence of excellence and market-oriented specialization, whether in science, engineering or business administration. This is precisely what critics mean by saying that the educational sector (as well as others) lacks flexibility. It is not the institutions that are inflexible, it is the politicians scared of "élitisme" who impose rigidity. And, of course, the insistence of the entitlement-oriented judicial system has created a situation in which incompetent professors cannot be denied tenure and further drag down the quality of university standards.[42]

The problem of flexibility worsens as one descends the educational scale. At the elementary and secondary levels it is not so much funding

---

40   Henry Ergas, "Does Technology Policy Matter?" in B.R. Guile and H. Brooks (eds.), *Technology and Global Industry*, Washington: National Academy Press, 1987, p. 233.

41   *Au courant, op.cit.*, p. 7.

42   The reader is reminded that singling out education for attention in the innovation context may amount to special pleading by the author, who is a university professor.

Figure 22: Operating Grants to Universities per FTE 1977-78 to 1990-91

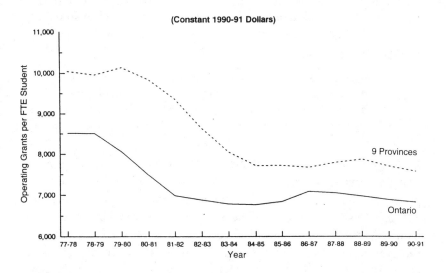

Figure 23: Growth in State and Provincial Grants Per FTE Enrolment, 1977-78 to 1989-90

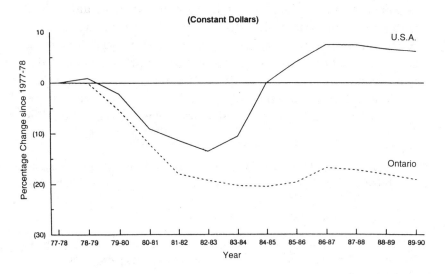

Source: Council of Ontario Universities, *The Financial Position of Universities in Ontario*, 1991, Toronto, n.d.

as incentives that are lacking. Our governments, so keen on promoting competition between businesses, neglect perhaps the most important business of all: the basic education of our young. Our schools have no bottom line and therefore no incentive to innovate and compete with each other. A voucher system could introduce such competition, but as yet this important innovation in education is far from the mainstream of policy discussion in Canada.

Investment in human capital or the educational part of an economic development infrastructure has been singled out for attention as merely one facet of the legitimate indirect or target-neutral policies that government ought to take care of. Deficiencies on a broader front in this respect have been discussed at length elsewhere, such as in the recent Economic Council's *Pulling Together*. Yet it is only fair to be reminded of Canada's federal government efforts since 1984 to reduce the deficit and to make the economy more competitive by opening up to international flows of trade and capital—and to be reminded to what extent its efforts in preparing Canada for the global economy are being frustrated by provincial prodigality and protectionism, often clothed in industrial policy justification.

Before we close it is useful to say a few words in support of government funding of basic research, of basic science. Such support is often difficult to justify, for there is no identifiable "optimal" level for the funding of basic research, nor can a single distribution pattern be identified as preferable to all others. Yet basic research, because it is deemed not to have appropriable results, qualifies as a traditional public good in need of community financing. While the federal government has not decreased its real dollar level of support to the three federal granting councils (National Science and Engineering Council, Medical Research Council, Social Sciences and Humanities Research Council),[43] moves are afoot to push basic research into more narrowly defined, "strategic" areas, to make it "less basic," more "capturable." This does not seem appropriate, even though at first sight the benefits of a national government's support to basic science could spill over to other countries. As Pavitt points out, an active national competence in basic

---

43   $430, $240 and $75 million, respectively in 1991-1992.

research is a necessary condition for benefiting from research undertaken elsewhere in the world and can be viewed as a national scientific
intelligence system.[44]

What is more, both basic research in the industrial firm and—what
is perhaps more apposite—in the university appears increasingly to
have substantial and measurable returns, private and social, respectively.[45] Václav Havel presented this position to a gathering of scientists
in the following manner:

> I would like to mention one kind of investment that has a
> singular character: at first sight it seems that demonstrably it
> does not and cannot yield returns, not to speak of profits, and
> so it is more or less money wasted. But it is only the first glance
> that says so. Once we think awhile about such investments, we
> notice that they are perhaps the most favourable investments of all.
>
> Yes, you suspect rightly that I speak of investments in educa-
> tion, science, research and culture. . . . Monies that society invests
> in these areas, whether in the form of grants from the budget, of
> tax relief, or by creating various foundations, do not really bring
> returns in the sense that an economist or accountant should be
> able to calculate what profit they would bring us tomorrow, the
> day after tomorrow, in a year or in five. . .
>
> As scientists you know very well how important is the so-called
> basic research. And this not only because no one ever knows in
> advance what practical consequences it can bring, but because
> of deeper reasons: the quest to understand the world is a funda-
> mental dimension of human existence; without exploration of
> the world a man is truly not human and every attempt to restrain
> human understanding with the shackles of practical applicabil-
> ity is in fact an attempt to enslave the human spirit. Basic
> research is understandably an expensive activity. But to pay for
> it means to pay for man being more in identity with himself.[46]

---

44 Keith Pavitt, "What Makes Basic Research Useful?" *Research Policy*, 1991,
    pp. 109-119.

45 Edwin Mansfield, "Academic Research and Industrial Innovation," *Re-
    search Policy*, 1991, pp. 1-12.

46 Václav Havel, "Investment in Education," *Vesmír*, 1991, No. 2, p. 64.
    Translated from Czech by the author.

Conclusion

Canadian governments spend more than adequate funds supporting activities designed to bring forth new knowledge, new products, and new processes as well as on activities which facilitate the diffusion of innovations.

Innovativeness, in both its facets of research and diffusion, is reasonably alive and well in this country. In large part this is due to the much-maligned presence of multinationals, domestic and foreign, who "invisibly" import research or diffuse its results.

The contribution of government subsidies to our innovative successes is not clear, but the evidence suggests it to be small in certain areas. Subsidies to enterprises are often simply substitutes for activities that businesses would have carried out anyway. Subsidies to basic research (usually carried out in universities) seem to be of more lasting value.

Perhaps what we have not stressed sufficiently is the most fundamental dimension of the competitive performance of a country's economy. This is the level of access of the final user, industrial buyer or consumer, to innovative products and processes in the market. The Canadian customer is, in this respect, more favoured than most others. The country is one of the most trade-intensive in the world, with increasing openness to the world's most innovative economy.

Thus, if only for these three reasons, there is no call to *intensify* direct policies supporting innovation. There is plenty of advice about how to improve the existing policies, some of it mentioned in this book.

What is needed is a major effort to enhance the quality of the business climate for innovation and competitiveness. This can be achieved by a return to sound fiscal policies and adequate infrastructure investment, at the federal and especially at the provincial levels. On how to do this more than enough sound advice is available, but not nearly enough readiness to accept at least some of it.